Indian Agriculture and Information and Communications Technology (ICT)

Indian Agriculture and Information and Communications Technology (ICT)

Edited by

M. Hilaria Soundari
Assistant Professor,
Department of Applied Research,
Gandhigram Rural Institute,
Gandhigram, Tamil Nadu

New Century Publications
New Delhi, India

NEW CENTURY PUBLICATIONS
4800/24, Bharat Ram Road,
Ansari Road, Daryaganj,
New Delhi – 110 002 (India)

Tel.: 011-2324 7798, 4358 7398, 6539 6605
Fax: 011-4101 7798
E-mail: indiatax@vsnl.com • info@newcenturypublications.com
www.newcenturypublications.com

Editorial office:
LG–7, Aakarshan Bhawan,
4754-57/23, Ansari Road, Daryaganj,
New Delhi – 110 002

Tel.: 011-4356 0919

First Published: **2011**

ISBN: **978-81-7708-254-8**

Published by New Century Publications and printed at Salasar Imaging Systems, New Delhi.

Designs: Patch Creative Unit, New Delhi.

PRINTED IN INDIA

About the Book

Farming is both a way of life and the principal means of livelihood for 65 percent of India's population of 110 crore. In spite of huge government machinery, Indian farmers still suffer from the absence of right information at the required time. The farmers require timely information on weather conditions, sowing time, availability of inputs including credit, expert advice on maintaining the crop in healthy condition, information on markets and other areas of interest to them and their families. In spite of the best efforts and expenditure, the conventional apparatus has not been able to deliver the goods satisfactorily. Herein lies the role of Information and Communications Technology (ICT) which can efficiently address the concerns of farmers stationed even at remote locations of the country. Low literacy levels, cost of computers, poor communications infrastructure make it impossible for individual farmers, particularly small farmers, to directly adopt ICT. This calls for institutional efforts to provide ICT-based services to farmers.

The present volume contains 11 well-researched papers by scholars in the field which provide deep insights into the various dimensions of the application of ICT in agriculture.

Editor's Profile

Dr. M. Hilaria Soundari is Assistant Professor in the Department of Applied Research, Gandhigram Rural Institute, Gandhigram, Tamil Nadu. She is a social activist, an academician and a researcher. She belongs to the Congregation of the Sisters of the Cross of Chavanod. She was awarded UGC-NET JRF/SRF and completed her doctoral degree in the Department of Applied Research, Gandhigram Rural Institute in 2003. Now, as a faculty member in the same department, she teaches courses on research methodology, development research and social work. She is an expert in participatory research techniques and computer applications in research including *SPSS and Ethnograph*. She is the principal investigator of UGC-Major Research Project, 'Achieving MDGs: Role of Marginalized Women and Micro-Credit'. She served the marginalized and downtrodden at the grassroots level for 12 years. She organized Self-Help Groups (SHGs) for Dalit women and children under an empowerment promoting agency, DEEPAM. She has published a number of scholarly articles in renowned national and international journals. She has also authored two books, viz. *Dalit Women: Dimensions of Empowerment* and *Emerging Dimensions in Self-help Groups*. She has participated and presented papers in international conferences held in USA, Netherlands, Switzerland, China and Germany.

Contents

Contributors and Their Affiliations

M. Hilaria Soundari	Assistant Professor, Department of Applied Research, Gandhigram Rural Institute, Gandhigram, Tamil Nadu.
K. Balasubramanian	Consultant, Common Wealth Learning, Canada.
R. Meenakshi	Reader, Department of Economics, Sri Sarada College for Women, Salem.
J. Gayathri	Ph.D. Scholar, Department of Economics, Sri Sarada College for Women, Salem.
A. Thomas	Lecturer (SGL), Arul Ananthar College, Karumathur, Madurai.
M.A. Sudhir	Professor and Head, Department of Applied Research, Gandhigram Rural Institute, Gandhigram, Tamil Nadu.
Ramasubramanian. M	Associate Professor, Department of Agricultural Extension and Rural Sociology, Agricultural College and Research Institute, Madurai.
Sujhi. G	Senior Research Fellow, Department of Agricultural Extension and Rural Sociology, Agricultural College and Research Institute, Madurai.
Rathakrishnan. T	Professor and Head, Department of Agricultural Extension and Rural Sociology, Agricultural College and Research Institute, Madurai.
Anandaraja. N	Assistant Professor, Department of Agricultural Extension and Rural Sociology, Agricultural College and Research Institute, Madurai.
Suganthi. N.S	Senior Research Fellow, Department of Agricultural Extension and Rural Sociology, Agricultural College and Research Institute, Madurai.
Poonthenmozhi. S	Senior Research Fellow, Department of Agricultural Extension and Rural Sociology, Agricultural College and Research Institute, Madurai.

Samwel Kakuko Lopoyetum	Senior Post Doctoral Research Fellow (UGC), Department of Co-operation, Faculty of Rural Social Science, Gandhigram Rural Institute, Gandhigram,, Tamil Nadu.
K. Rakes Sarma	Department of Geoinformatics, Faculty of Rural Development, Gandhigram Rural Institute, Gandhigram, Tamil Nadu.
Rangamlian Aimol	Assistant Professor, Faculty of Rural Development, Gandhigram Rural University, Gandhigram, Tamil Nadu.
P. Mooventhan	Senior Research Fellow, Department of Agricultural Extension and Rural Sociology, Agricultural College and Research Institute, Madurai.
H. Philip	Professor, Educational Media Centre, Directorate of Open and Distance Learning, Tamil Nadu Agricultural University, Coimbatore.
A. Thomas William	Reader, Arul Ananthar College, Karumathur, Madurai.
G. Albin Joseph	Research Scholar, Arul Ananthar College, Karumathur, Madurai.
P. Mary Elizabeth	Scholar, Department of Sociology, Gandhigram Rural University, Gandhigram, Tamil Nadu.

Editor's Note on the Book

At the time of India's Independence, Jawaharlal Nehru said, "Everything else can wait, but not agriculture". Unfortunately this profound truth is yet to be converted into concrete policies and action on an adequate scale. There have been several policy statements for agriculture during the last 60 years, including the comprehensive reports of the National Commission on Agriculture, 1976. The last National Agricultural Policy was formulated in 2002. In fact, agricultural growth has decelerated during the last decade. This has led to a decline in real per capita incomes in rural India, in comparison to the rapid growth in urban incomes. The National Policy for Farmers, 2007 is designed to fill this gap and to get the focus of agricultural policies shifted to the women and men feeding the nation. Its main aim is to bring about a mindset change and a shift from an approach of patronage to partnership with farm and fisher families based on mutual respect. Progress in agriculture should be measured by the growth rate in the net income of farm families, if the human dimension is to be added to agricultural policies.

Farming is both a way of life and the principal means of livelihood for 65 percent of India's population of 110 crore. India's farm population is increasing annually by 1.84 percent. The average farm size is becoming smaller each year and the cost-risk-return structure of farming is becoming adverse, with the result that farmers are getting increasingly indebted. Marketing infrastructure is generally poor, particularly in perishable commodities. No wonder, a recent NSSO survey revealed that nearly 40 percent of farmers would like to quit farming, if they have the option to do so. Unfortunately, there is little option for them except moving into urban slums. The support systems needed by farmers, like research, extension, input supply and opportunities for assured and remunerative marketing are in need of review and reform. Small farmers are forced to borrow from moneylenders at high rates of interest, since only 51 percent of the credit requirements of farmers are met by institutional sources. Only 27 percent of all cultivator households receive institutional credit.

Policies are needed for making sub-marginal, marginal and small farmers economically viable and environmentally sustainable. Technology has been a major factor in the rich-poor divide until now. We should now enlist technology as an ally in the movement for gender and social equity. This will call for a pro-poor, pro-women and pro-nature orientation to technology development and dissemination. Also, Intellectual Property Rights (IPR) policies should be such that there is social inclusion in access to technologies. This will call for a considerable stepping up of investment in public good research and an *antyodaya* approach to technology development and dissemination. There is no option except to produce more food and other commodities under conditions of diminishing per capita arable land and irrigation water resources. Hence, we must harness the best in frontier technologies and integrate them with traditional wisdom and thereby launch an eco-technology movement.

Information and Communications Technology (ICT) has been an engine of growth in various sectors. Certainly it can effectively be used to enhance the livelihood of farmers. These changes can radically lead the country in to a new era.

In spite of huge government machinery, Indian farmers still suffer from the absence of right information at the required time. The farmers require timely information on weather conditions, sowing time, availability of inputs including credit, expert advice on maintaining his crop in healthy condition, information on markets and on all other areas of interest to him and his family. In spite of the best efforts and expenditure, the conventional apparatus has not been able to deliver the goods satisfactorily. Herein lies the role of ICT which can efficiently address the concerns of farmers stationed at even remote locations. Low literacy levels, cost of computers, poor communication infrastructure make it impossible for individual farmers, particularly small farmers, to directly adopt ICT. This calls for institutional efforts to provide ICT-based services to farmers.

Over the last two decades, India has emerged as the most preferred destination for outsourcing of ICT services. The vibrant ICT industry is contributing immensely by providing information about latest technology and international business practices. Various sectors of the Indian economy such as industry, finance, insurance, communications and transport have adopted information technology in a big way. However, agricultural sector of the economy is lagging behind in utilizing ICT services. Some efforts have been made in this regard but they are by no means adequate. The rapid strides that the country has registered in the ICT field will remain incomplete unless ICT is fully utilised to ensure more efficient and productive Indian agriculture. Practically there is no area in agriculture in which ICT has no role to play. The immense scale and diversity of Indian agriculture provides the ultimate challenge to the potential of information technology.

The adoption of ICT services in agriculture depends on the main functionaries involved in Indian agriculture. These functionaries can make a big contribution to the growth of agriculture with the assistance of ICT. For this purpose, the following functionaries can be considered (a) farmers, (b) industries providing inputs to agriculture, (c) industries dealing with agricultural output, (d) central and state governments and (e) NGOs working for the benefit of farmers and agricultural universities and research centres.

In order to keep pace with the state-of-the-art technologies, NIC has been conducting various training programmes on ICT application on a regular basis from time to time for the user organisations of agriculture sector. Agricultural Informatics Division of NIC has taken up various initiatives in bringing ICT led development which includes Web enabled applications, GIS based applications, Multimedia applications, Database applications and e-Governance and training etc. in the Ministry of Agriculture.

The face of the Indian agriculture can be transformed by a well conceived deployment of ICT. The potential of ICT as yet remains untapped and urgent measures are required to derive maximum benefit. The key players involved in this process such as industry, government and educational institutions and research centres are required to make contributions in this endeavour. The initiative to develop necessary ICT-based agricultural services need to be developed immediately.

Proper planning, monitoring and corrective action requires availability of timely information of right quality. The Central and State Governments require to hasten the process of computerisation of all the records and transactions. Since the government has a major say in promoting agriculture, this will have cascading effect on the entire rural areas. To begin with, it is necessary that the government agencies should accept conventional reports in the electronic format. This will bring about much desired speed and effectiveness in governance. Many State Governments are showing interest in developing proper systems for e-governance. But the implementation of the same is still requires a lot to be desired. The Central and State Governments should also make information available for ready reference by those interested. This will be not only helpful to people but also will raise interest levels in ICT-based services. Similarly, research institutions and NGOs should exchange and provide information through a well conceived network. These measures will encourage end user, i.e. farmer's participation and will provide the much needed input from them.

The present volume contains 11 well-researched papers by scholars in the field which provide deep insights into the various dimensions of the application of ICT in agriculture. I place on record my deep sense of gratitude to the distinguished authors for their worthy contributions and co-operation.

Gandhigram **M. Hilaria Soundari**
September 2010

1

Introduction

M. Hilaria Soundari

ICT Services for Indian Farmers

In spite of huge government machinery, Indian farmers still suffer from the absence of right information at the required time. The farmers require timely information on weather conditions, sowing time, availability of inputs including credit, expert advice on maintaining his crop in healthy condition, information on markets and on all other areas of interest to him and his family. In spite of the best efforts and expenditure, the conventional apparatus has not been able to deliver the goods satisfactorily. Herein lies the role of ICT which can efficiently address the concerns of farmers stationed at even remote locations. Low literacy levels, cost of computers, poor communication infrastructure make it impossible for individual farmers, particularly small farmers, to directly adopt ICT. This calls for institutional efforts to provide ICT-based services to farmers.

Over the last two decades, India has emerged as the most preferred destination for outsourcing of ICT services. The vibrant ICT industry is contributing immensely by providing information about latest technology and international business practices. Various sectors of the Indian economy such as industry, finance, insurance, communications and transport have adopted information technology in a big way. However, agricultural sector of the economy is lagging behind in utilizing ICT services. Some efforts have been made in this regard but they are by no means adequate. The rapid strides that the country has registered in the ICT field will remain incomplete unless ICT is fully utilised to ensure more efficient and productive Indian agriculture. Practically there is no area in agriculture in which ICT has no role to play. The immense scale and diversity of Indian agriculture provides the ultimate challenge to the potential of information technology.

The adoption of ICT services in agriculture depends on the main functionaries involved in Indian agriculture. These functionaries can make a big contribution to the growth of agriculture with the assistance of ICT. For this purpose, the following functionaries can be considered (a) farmers, (b) industries providing inputs to agriculture, (c) industries dealing with agricultural output, (d) central and state governments and (e) NGOs working for the benefit of farmers and agricultural universities and research centres.

In order to keep pace with the state-of-the-art technologies, NIC has been conducting various training programmes on ICT application on a regular basis from time to time for the user organisations of agriculture sector. Agricultural Informatics Division of NIC has taken up various initiatives in bringing ICT led development which includes Web enabled applications, GIS based applications, Multimedia applications, Database applications and e-Governance and training etc. in the Ministry of Agriculture.

National E-governance Plan (2003-07)

The national e-governance plan (2003-07) reflects the strategic intent of the central government in the right perspective. Many projects are earmarked under this plan, and it is trying to address the digital divide. From a political perspective, after watching the performance of some IT-savvy states in the recent elections, the system has woken up to the need to focus more on rural development. The political systems are keener to use IT to disseminate information faster to farmers, disburse loans, improve education and the health systems in villages etc. There is a clear-cut incentive to do it as 60 percent of the vote-bank still lives in rural India.

E-governance has to be supported by the will and

resources of those who are in governance, be it at the central or state level. The central government has analysed and appreciated the concept by creating a separate E-governance department headed by a secretary to trigger e-governance in India. The World Bank, ADB and UN have been approached, and in response they are generously funding e-governance projects. In future, education, agriculture, state wide area networks (SWANs) and Community Information Centre projects will be rolled out backed by a strong public private participation model (PPP) to achieve long-term sustainability. Projects with PPP models in these segments can revolutionise the governance experience.

Cyber Dhabas

To take ICT to the door steps of the farmers, it is proposed that a large number of access points called cyber dhabas are setup. These cyber dhabas should be provided with a PC and a modem with telephone connection to connect to the databases. The farmers can take the assistance of the operator of cyber dhabas for a nominal charge. The major bottleneck in spreading e-culture to rural areas is related to connectivity i.e. ensuring that the access points can get connected to the databases which are in selected locations. Since dialup lines are very slow, other viable options are required to be explored. Unfortunately, the alternatives are expensive and may not be feasible. One silver lining is that several private operators are connecting important cities with fibre optic which provides a very reliable and fast access. Since these will pass through rural areas, it is possible to explore the possibility of tapping this potential by laying the last mile connectivity. Once this is done, substantial segment of rural India can access the IT-based services.

E-Choupal

E-Choupal, ITC's unique web-based initiative, offers farmers the information, products and services they need to enhance productivity, improve farm-gate price realisation, and cut transaction costs. Farmers can access the latest local and global information on weather, scientific farming practices, as well as market prices at the village itself through this web portal-all in Hindi. E-Choupal also facilitates the supply of high quality farm inputs as well as the purchase of commodities at the farm.

Given the literacy and infrastructure constraints at the village level, this model is designed to provide physical service support through a choupal sanchalak, himself a lead farmer, who acts as the interface between the system and the farmers. The contents of this site in their entirety are made available only to the registered sanchalaks.

IFFCO-ISRO GIS Project

Recently, IFFCO has initiated ICT-based services for farmers and cooperative societies. It has taken up a project in association with Indian Space Research Organisation (ISRO) to utilise satellite based remote sensing data and Geographical Information Systems (GIS). Developed countries have been utilising precision farming with the help of ICT tools for a long time. While this will take a long time for our country due to small holdings, it is to be noted that GIS has an invaluable role to play even in the existing conditions. Remote sensing and GIS information can provide warnings on evolving crop stresses, crop vigour etc.

The IFFCO-ISRO GIS project extends support for efficient and timely availability of IFFCO's fertiliser to farmers though better logistics and efficient operations. Based on the experience gained from this project, more intensive services based on GIS are envisaged for the entire country. In addition to the GIS-based services, effort is being made to create databases that contain information of interest to the farmers. These include recommendation on package of practices for major cereals, pulses, horticulture, floriculture and animal husbandry etc. Information on various inputs such as seeds, fertilisers, credit is provided. An important service envisaged is to provide access to the nearest expert in case of stress or any other problem witnessed in the crops. Facilities are sought to be provided to encourage and share farm experiences by forging various crop forums. Many of the agricultural extension services are also proposed to be made online using multimedia facilities.

In order to encourage farmers to obtain best possible price, information regarding prices of various crops prevailing in approachable markets (mandies) is also being provided. Other areas of interest to farmers such as distance education, location specific news etc. are also planned.

Agricultural Marketing Information Network (AGMARKNET)

Market information is needed by farmers in planning production and marketing, and equally needed by other market participants in arriving at optimal trading decisions. The existence and dissemination of complete and accurate marketing information is the key to achieve both operational and pricing efficiency in the marketing system. Advancement in Information and Communication Technology (ICT) has made the world a smaller place and a larger market at one go. To fully utilize the new emerging trade opportunities for the benefit of farming community, there is a felt need to establish an ICT-based Agricultural Marketing Information Network in the country. Towards this end, Ministry of Agriculture, Government of India launched the ICT-based Central sector scheme of Agricultural Marketing Information Network (AGMARKNET) in March 2000 to link important agricultural produce markets spread all over the country and the State Agriculture Marketing Boards and Directorates. The project is being executed with the technical support of National Informatics Centre (NIC).

In order to facilitate wider market access to farm produce and better price discovery through value addition, the scheme is additionally disseminating information on market requirements in terms of quality, packing, standards, sanitary and phyto-sanitary conditions etc.

Objectives of the scheme are as under:

1. To establish a nation-wide information network for speedy collection and dissemination of market information and data for its efficient and timely utilization.
2. To facilitate collection and dissemination of information related to better price realization by the farmers. This would cover:

- Market-related information such as market fee, market charges, costs, method of sale, payment, weighment, handling, market functionaries, development programmes, market laws, dispute settlement mechanism, composition of market committees, income and expenditure etc.
- Price-related information such as minimum, maximum and modal prices of varieties and qualities transacted, total arrivals and dispatches with destination, marketing costs and margins etc.
- Infrastructure-related information comprising facilities and services available to the farmers with regard to storage and warehousing, cold storage, direct marketing, contract farming, buy-back arrangements, grading, re-handling and repacking etc.
- Promotion-related information covering accepted standards and grades, labelling, sanitary and phyto-sanitary requirements, pledge finance, marketing credit and new opportunities available in respect of better marketing.

3. To sensitize and orient farmers to respond to new challenges in agricultural marketing by using ICT as a vehicle of extension.
4. To improve efficiency in agricultural marketing through regular training and extension for reaching region-specific farmers in their own language.
5. To provide assistance for marketing research to generate

marketing information for its dissemination to farmers and other marketing functionaries at grass-root level to create an ambience of good marketing practices in the country.

The Scheme will continue during the Eleventh Five Year Plan (2007-12) by enhancing its scope to cover 500 additional markets including remaining principal regulated markets, if any, farmer/consumer markets, private markets, panchayat markets, urban markets and special commodity markets. Following activities are also proposed to be undertaken while implementing the Scheme during the Eleventh Plan to fully achieve various objectives of the Scheme:

1. Information regarding export documentation, export infrastructure, transportation facilities, exporters, marketing experts, agri-business corporate, quality standards, grading/packhouse infrastructure, plant quarantine facilities, package of best marketing practices, seasonality of commodities will be provided on the portal.
2. Information on contract farming, direct marketing opportunities, value addition facilities etc. to be enriched and tie-ups between farmers/growers and sponsors/buyers will be facilitated through the portal by evolving a specific software system through outsourcing of work to an expert agency. It would be developed as an interactive query based system to educate the farmers on issues raised by them and would serve as a national clearing house for buyback and contract farming arrangements.
3. Information on warehouses along with addresses, accreditation facilities and their distance, pledge financing facilities etc. will also be provided on the portal.
4. Area specific success stories will be loaded at the portal so that farmers get knowledge and can plan for trading of their produce.
5. GIS-based Atlas will be expedited for the content enrichment and system will be put in place for regular data updating.
6. A data validation software will be developed to identify

the wrong reporting and for removal of undesirable data from the central data base.

7. Electronic display boards will be provided at every networked market for displaying minimum and maximum price of important commodities, arrivals, temperature etc. These efforts would be dovetailed, as far as possible, with the action being taken by the Forward Market Commission under the Scheme of the Department of Consumer Affairs.
8. Translation of contents of portal into regional languages is necessary and this work will be undertaken through the concerned State Marketing Boards/Directorates.
9. Strategic alliances will be developed with corporate, telecom players and private users for strengthening market intelligence services.

Promotion of nationally and internationally acceptable standards of grading and standardization, packaging and labelling, storage and warehousing and sanitary and phyto-sanitary requirements and quality certification in farm sector will enable trade and processing sector to undertake large scale agricultural marketing operations in domestic as well as international markets.

Local Area Network (LAN) at Ministry of Agriculture

A high speed Local Area Network has been established in the Ministry of Agriculture spread over various buildings viz. Krishi Bhawan (~500 nodes), Shastri Bhawan (~200 nodes), Krishi Anusandhan Bhawan (~100 nodes) etc. In addition, Agricultural Informatics Division has extended its full support (round the clock operation) to the Ministry of Agriculture at the time of super Cyclone in Orissa and Gujarat Earthquake. Video Conferencing facilities were used extensively from the Krishi Bhawan studio to Orissa and Gujarat on continuous basis.

The Central sector scheme "DACNET" has been approved by the Department of Agriculture and Co-operation to be implemented by NIC. The project will commence in this financial year. The project proposal envisages establishment of NICNET based INTRANET linking directorates, attached offices, subordinate offices, autonomous bodies and public sector undertakings and field units of the Department of Agriculture and Cooperation, Ministry of Agriculture. This project includes networking of field offices of DAC, connectivity, procurement of H/W, S/W tools, Application Software Development and training of officials.

Plant Protection Informatics and Communication Network (PPIN) envisages to link the Directorate of Plant Protection Quarantine and Storage (DPPQS), Faridabad, National Plant Protection Training Institute, Hyderabad, CIMPS, LCC, Central Pesticides Testing Laboratories, and State Pesticides Testing Laboratories and to disseminate information related to plant protection and pesticides. As a part of project, the computerization of the CIB/RC of DPPQS has been taken up by NIC.

To cater to the needs of the Department, NIC has developed various information systems viz. FPO and licensing monitoring, Plan scheme monitoring, industrial approvals, processed food products, FPO on-line status, industrial entrepreneur's memoranda query information, country profile and international cooperation etc.

Agricultural Extension Information System Network: VISTARNET: Research, education, extension, and training are considered as four pillars of sustainable agriculture. Generation and transfer of technology have become very crucial the world over. As a step towards making technology reach the small holders (resource-poor-farmers), efforts are being made to establish "VISTARNET–NICNET based Agricultural Extension Information System Network, in India, linking extension functionaries at Central, State, and District level. The required funds for implementing VISTARNET will be provided through the National Agricultural Technology Project (NATP). As a part of VISTARNET, Informatics Development for the Directorate of Extension has been taken up by NIC.

Indian Agriculture On-line: NIC has submitted an IT-Plan for the agriculture sector with the objective of establishing NICNET based Agricultural Informatics and Communication Network (AGRISNET) to facilitate higher and sustainable agriculture productivity and also putting Indian Agriculture on-line in the Country.

Animal Production and Health Information Network (APHNET): APHNET envisages building up of reliable databases and network based information systems for all activities of the animal husbandry and dairying sector at district, state and national level, using NICNET facilities. In view of importance of making technology reach the small holders through IT, the plan also suggests to establish APHNET nodes at about 42,000 veterinary polyclinics, hospitals, and dispensaries and veterinary aid centres for strengthening animal disease surveillance and advisory system in the country. As a part of APHNET, National Project on Rinderpest Eradication (NPRE) which aims at capturing animal disease related information from various State Animal Husbandry Departments has been entrusted to NIC.

Agricultural Research Information System Network (ARISNET): The ARISNET, a close user group of institutions in National Agricultural Research System of India (NARS) on NICNET, links ICAR Institutions and their Regional Research Stations, Central Agricultural University, State Agricultural Universities and their colleges, Krishi Vigyan Kendras, Zonal Research Centres and Project implementing centres with the ICAR Headquarters. ARISNET has become an integral part of agricultural research, extension and education process.

Infrastructural Facilities and Services Provided by NIC to the Ministry of Agriculture

These are the following:

1. INTRANET/INTERNET.
2. Telecommuting Programme.

3. Video Conferencing.
4. Information Kiosk.
5. In-house Training facilities.
6. Web site design and development.
7. Web enabled applications.
8. Information Bulletin.
9. Intranet applications.
10. Agriculture Portal.
11. NICNET based Public Information and Facilitation Centre.

Market Information System for Horticulture

This project has been taken up by NIC on a turnkey basis. As part of the project, NICNET based internet/intranet facilities have been established at National Horticulture Board (NHB), Gurgaon. Computing facilities have been created at 33 market centres of the Board located all over the country. NIC has developed and implemented the necessary software for evolving a comprehensive database of the prices and arrivals of fruits and vegetables being received by NHB headquarters from 33 market centres on a daily basis. This information is being used by the Ministry of Agriculture, its related agencies and markets through National Bulletin on fruits and vegetables.

Integrated Fertilizer Management Information System (FERMIS)

The Department of Fertilizers (DOF) in collaboration with National Informatics Centre (NIC) has introduced computer-based methods for decision support with the major objective to evolve an evaluation system which ensures a uniform system of Planning and control mechanism with signalling system to highlight deviations from desired performance indicators by Plants/organisations for all the public sector fertilizer enterprises. Thus, an Integrated Fertilizer Management Information System (IFMIS) of worth Rs. 1.99 crore has been executed through NIC and various IT based systems developed covering Planning, Movement, Import, Handling-Payment, Project Monitoring, Performance Monitoring and Evaluation aspects to strengthen Fertilizer Informatics for decision support. Information exchange for decision support has been promoted through Fertilizer Informatics Network (FERTNET) extended to Department of Fertilizers and various Fertilizer Companies viz. IFFCO, NFL, MFL, GNFC, FCI and HFCL.

Computerisation of Agricultural Census and Input Survey

Recognizing the predominance of the agriculture sector in the Indian economy, Department of Agriculture and Cooperation collects and maintains agricultural statistics such as number, area, tenancy, land utilisation, cropping pattern and irrigation particulars of different classes of operational holdings regularly and make it timely accessible to the planners and policy makers for decision making. This project was entrusted to NIC and it made significant headway in almost all the States/UTs except Bihar, where Census is yet to be conducted. The salient features are the following:

1. Creation of large database of about 8000 Million Bytes at National Level and 1 GB at state level.
2. Processing and tabulation of Agricultural Census and Input Survey database at District/State/National level.
3. Development and implementation of information retrieval system at micro and macro level for decision making at various levels.
4. Selective information will be put on INTERNET/INTRANET for easy access of information to public to ensure transparency in the Government functioning.
5. Training programmes on software implementation for officers of States/UTs and Central Government.

Summing Up

The face of the Indian agriculture can be transformed by a well conceived deployment of IT. The potential of IT as yet remains untapped and urgent measures are required to derive maximum benefit. The key players involved in this process such as industry, government and educational institutions and research centres are required to make contributions in this endeavour. The initiative to develop necessary IT-based agricultural services need to be developed immediately. Parallel steps to develop necessary IT communication infrastructure are to be taken up along with the utilisation of fibre optic network wherever it is passing through the rural segments.

Proper planning, monitoring and corrective action requires availability of timely information of right quality. The Central and State Governments require to hasten the process of computerisation of all the records and transactions. Since the government has a major say in promoting agriculture, this will have cascading effect on the entire rural areas. To begin with, it is necessary that the government agencies should accept conventional reports in the electronic format. This will bring about much desired speed and effectiveness in governance. Many State Governments are showing interest in developing proper systems for e-governance. But the implementation of the same is still requires a lot to be desired. The Central and State Governments should also make information available for ready reference by those interested. This will be not only helpful to people but also will raise interest levels in IT-based services. Similarly, research institutions and NGOs should exchange and provide information through a well conceived network. These measures will encourage end user, i.e. farmer's participation and will provide the much needed input from them.

The Bhoomi Project of Karnataka Government has led to the computerisation of the centuries-old system of handwritten rural land records. Through it, the revenue department has done away with the corruption-ridden system that involved bribing village accountants to procure land records; records of right, tenancy and cultivation certificates. The project is expected to benefit seventy lakh villagers in 30,000 villages. A farmer can walk into the nearest taluk office and ask for a computer printout of his land record certificate for Rs 15. He can also check details of land records on a touch-screen kiosk by inserting a two-rupee coin. These kiosks, installed at the taluk office, will provide the public with a convenient interface to the land records centre.

2

Village Knowledge Centres (VKCs) and Open/Distance Learning

K. Balasubramanian

The differences between information and knowledge are being spelt out in many books and papers in recent times. Many authors have described the progressive processes from data to information to knowledge to wisdom in terms of purposes and contexts. Data refers to raw materials such as facts and figures that could be collected by an information system. Information refers to analyzed data often presented in a form that is specifically designed for a given decision-making task, and transmitted to/received by decision makers. Knowledge refers to subsequent absorption, assimilation, understanding and appreciation of that information (Chapman and Slaymaker, 2002). Pomeroll and Brezillon (2001) quoting Newell and Simon (1972) argue that knowledge is information incorporated in an agent's reasoning and made ready either for active use within a decision process or for action. It is the output of a learning process. Thus the roles of knowledge are to: (1) to transform data into information, (2) derive new information from existing ones, and (3) acquire new knowledge pieces. Wisdom is considered as meta-knowledge, knowledge mobilized to acquire new knowledge and update it. From a philosophical angle wisdom refers to the evaluation of knowledge vis-à-vis the norms, values and morality (Pomeroll and Brezillon, 2001).

Knowledge management focuses on definition of the context and validation of the information. It also increases the connections among people (who have knowledge) that would likely not occur without the help of a knowledge management system (Terra and Angeloni, 2002). They have identified the process of searching answers for the following questions as an important dimension of knowledge management:

- Who created the information?
- What is the background of the creators of information?
- Where and when was it created?
- How long will the information be relevant, valid and accurate?
- Who validated the information?
- Who else might be interested or has similar knowledge?
- Where was it applied or proved to be useful?
- What other sources of information are closely related?
- How to test and validate some of the concepts?

Agricultural Extension and ICT

In the context of rural community, the presence of traditional knowledge is another important dimension of knowledge management. The social construction of traditional knowledge and the blending of the new knowledge with traditional knowledge are the components of knowledge management. Thus knowledge management necessitates a participatory management in which the rural community plays a crucial role of absorption, validation, critical evaluation, assimilation, understanding and appreciation of information. A paradigm shift in the concept and practices of extension will occur only when the community develops its own framework for knowledge management.

In many projects connected with agricultural extension and Open and Distance Learning (ODL), modern information and communication technology in the form of tele-centres is being given utmost importance. A rural tele-centre evolves into a knowledge centre only when modern ICT facilitates transfer of information into knowledge. A knowledge centre is important for strengthening the self-directed learning which is an important learning process in the human society. A tele-centre providing market price is an information centre. A tele-centre, which enables the rural community to understand the differential mechanisms through which prices are influenced and determined, is a knowledge centre.

Most of the rural ICT projects focus on providing information. In agriculture and rural development, the importance of uneven distribution of knowledge in explaining variations in Total Factor Productivity (TFP) is being increasingly recognized (Chapman and Slaymaker, 2002). Mere information in the form of flow of messages may not be able to address the problem. Knowledge as the creative result of a flow of messages anchored on the commitment and beliefs of the actors involved in the process and resulting in human action is needed. Environment in which knowledge is built; capacity building and empowerment processes, social mobilization and organization are the important factors which to be taken into consideration while transforming a tele-centre into a knowledge centre. Freire (1973) argued in the case of the pedagogy of oppressed, the need for dialogues and discourses among learners to understand the world instead of mere understanding of words. Similarly in the process of knowledge management, dialogues and discourses of among the rural community are essential. Modern ICT, if properly defined can help to broaden the canvass for dialogues and discourses among the rural community.

Community and ICT

The evolution of an active utilizer constituency is the basic premise for a strong extension framework. Such a constituency will help to provide services to its members in time and space through horizontal transfer of knowledge. According to Suliaman and Holt (2002), 'the poor face substantial transaction cost in accessing means of production, in adding value to their produce and in accessing markets for it. The core of extension is in helping people make better choices through the supply of information and in enhancement of people's capacities to process such information and act on it, and thereby reducing the transaction costs involved in pursing livelihood options.'

The development of active utilizer constituency and horizontal transfer of knowledge have become crucial when particularly viewed in the context of declining strengths of the resources in the formal governmental extension agencies and the increasing demand from the rural sectors. Formal extension would become fruitful if it becomes a facilitating process for active utilizer constituency and community knowledge management, rather than attempting direct interventions and implementations. Such a framework requires a capacity building process among the officials of the formal extension system in building the structures, shaping the functions and refining the process through mobilization, organization, technology incubation and systems management.

Capacity building through Open and Distance Learning (ODL) particularly ICT based ODL, in agricultural sector is being emphasized by many national and international organizations. It is important to realize that farmers and agricultural labourers should not be treated as mere consumers of generic information. ODL should not be seen in isolation. It should become a part of total extension package in which horizontal transfer of knowledge, participatory research and knowledge management play a crucial role. Three important characteristics differentiate the "new ODL" approach with the conventional ODL approach:

- Unlike the conventional ODL which is related to individual learners, the new ODL in agricultural sector requires a well organized learning community in the form of farmers association, cooperatives, SHGs, women's groups etc.
- ODL should be a part of a blended learning process in which face to face interaction, learning by doing, learning through evaluation and experience, participatory research etc.
- The rural community should not be seen as a mere

consumer of knowledge and information. The ODL should be structured in such a manner through which the rural community is able to convert the generic information into location specific knowledge and then empower its members through horizontal transfer of knowledge. It should enhance the self-directed learning among the rural community.

References

Chapman, Robert and Tom Slaymaker (2002). "*ICTs and Rural Development: Review of the Literature, Current Interventions and Opportunities for Action,* ODI Working Paper 192.

Freire, P. (1972). "*Pedagogy of the Oppressed*", Harmondsworth: Penguin.

Newell A. and Simon H. (1972). "*Human Problem Solving*", Prentice-Hall, Englewood Cliff.

Pomerol, J. and P. Brézillon (2001). "About some relationships between Knowledge and Context". http://www-poleia.lip6.fr/~brezil/Pages2/Publications/CXT01/JCP-PB.pdf.

Roling, Neils (1988). "*Extension Science, Information System in Agricultural Development*". Cambridge University Press: Cambridge.

Shenggen Fan (2002) http://poverty.worldbank.org/files/11501_PresentationSFan.pdf.

Sood, Aditya Dev (2003). "*Information Nodes in the Rural Landscapes* May-June" www.i4donline.net.

Suliaman Rasheed,V., Georgina Holt (2002). "*Extension, Poverty and Vulnerability in India*", Working Paper 154, ODI: London

Terra, J.C., Terezinha Angeloni, (2002). "*Understanding the difference between Information Management and Knowledge Management*" www.kmadvantage.com/docs/km_articles/Understanding_the_Difference_Between_IM_and_KM.pdf.

Whyte, Anne (2003). "*Understanding the Role of Community Telecentres I Development-A Proposed Approach to Evaluation*" http://www.idrc.ca/telecentre/evaluation/html/30a_Und.html.

3

ICT and Sustainable Agriculture Development

R. Meenakshi and J. Gayathri

Introduction

Farmers in India operate in widely distributed rural locations, which deprive them of the benefits of timely and useful information from research centres and service organizations. A comparison of the yield of rice, cotton, Soya beans and maize between India and China or Egypt, indicated that Indian generates only a third of what is produced elsewhere. This is because of the inadequate dissemination of agricultural technology and its adoption. Hence, information technology (IT) has a definite role to play in making the Indian farmer more prosperous.

Practically, there is no area in agriculture in which IT has no role to play. In the light of Globalization and the impact of WTO on agriculture and agro-based industries, the use of IT based services in this field has become imperative. The immense scale and diversity of Indian agriculture provide the ultimate challenge to the potential of information technology. The rapid strides that the country has registered in the IT field will remain incomplete unless IT is fully utilized to ensure more efficient and productive Indian agriculture.

Since IT has revolutionized the world and forms an integral part of all sectors, e-commerce has a tremendous potential in agribusiness too. Information of the required quality has the potential of improving efficiency in agriculture. Hence, there is a need to examine the latest developments in IT that facilitate effective IT penetration into rural areas and types of systems required in this post-WTO environment. With modern IT, extension services can provide advice tailor-made for each farmer to help him select the most profitable route to creating optimal conditions for plant growth.

Due to population growth and the shrinking of natural resources, the increase in food productivity has to be through higher and more efficient use of inputs, like quality seeds, fertilizers and water. The availability of timely and adequate information on these critical inputs is essential for sustainable high productivity and food security. This depends on the judicious use of natural resources, such as soil, water, live stocks, plant genetics, forestry, climate, rainfall and topography. Developments in enabling technologies, such as information and communication technology (ICT), geographical information system technology (GIST) and global positioning system technology (GPST) facilitate informatics- led agricultural development, which will be a step towards enhancing the quality of life of the farming community.

Agriculture has undergone a number of changes in the 20th century, including new farm machinery, intensive fertilizers and agro-chemical management, plant breeding, high-yielding varieties of seeds and genetics manipulations. The growing demand for foodgrains, vegetables, fruit, milk, poultry and newer challenges to agriculture, ICT, bio-technology and environment technology are viewed as the drivers of globalization in the scenario of liberalization, privatization and a tighter Intellectual Property Rights regime. Moreover, the global environment arising out of WTO, poses challenges requiring Indian agriculture to be competitive, sustainable, viable and equitable. In addition, until harvest, agriculture is a state subject, while afterwards it becomes a commodity of global trade. Hence, it is all the more important that agriculture be given its due recognition.

India is vast and has a variety of landforms, climate, geology, physiography and vegetation. It has great regional diversity, which accounts for its uneven 'economic and agricultural' development. India is becoming an information society with emphasis on IT. The fusion of various technologies aided by IT will bring in the desired inputs for sustainable agricultural and rural development. Hence, these technologies will have many potential applications spanning the use of agriculture in all walks of life- from the farmer to the cooperative and professional bodies, from farm machinery vendors, fertilizer and chemical companies, insurance and commodities to agronomists, consultants and farm advisors.

Since farming is also a business, the farmer as a businessman takes several decisions to maximize returns. It can be used to meet most of his information and commercial needs. Technology dissemination is crucial for getting across the benefits of any technology to farmers. But traditional methods are not cost effective. So a central repository for storing information in local languages is required.

ICT and the Farmers

The basic information generated by various agencies must be analysed and presented to the farmer with respect to location specific crops, forecasts of pests, diseases, impact of weather on crops, etc. An analysis of market-related information, namely the impact of cropping patterns and effects of exports/imports on prices, etc, must also be done.

The differences between farming practices in developed countries and in developing countries are reflected in the developed and underdeveloped areas within the same country also. In addition, the ground realities often do not reflect the results of research, due to an awareness gap, which can be filled by an IT revolution to stimulate agricultural development. This can give small farmers the flexibility they need to change crop choices, develop products for small markets and market directly to distantly located consumers, too. Organisations of small producers can strategically plan for low-cost inputs, better storage facilities, improved transportation links and collective negotiations with buyers for better remuneration in order to directly benefit the farmers.

In the context of agriculture, the potential of IT can be assessed under two heads:

- As a direct tool for contribution to agricultural productivity;
- As an indirect tool for empowering farmers to take decisions that have a positive impact on agriculture, known as agricultural extension services.

Precision farming extensively uses IT to make a direct contribution to agricultural productivity. The technique of remote sensing using satellite technologies, geographical information systems, and agronomy and soil sciences, are used to increase agricultural output, which are more suitable for corporate farming. However, the indirect benefits of IT in empowering Indian farmers require timely and reliable sources of information inputs for taking decisions. The changing environment faced by Indian farmers makes information not merely useful, but necessary so that they can remain competitive.

The emerging scenario of a deregulated agricultural system has made the deployment of IT necessary as a strategic tool for the benefit of rural India. These facilitate a proper understanding of the implications of the WTO on Indian agriculture, without any ambiguity. The mandatory changes in government policies on tariffs, imports, year-wise phasing of the same and the impact on various subsidy schemes, would be of concern to people. Since the removal of restrictions on Indian agricultural markets, the macro economic situation related to foreign exchange, inflation etc, has a significant impact on Indian agriculture.

Decision Support Systems for Farmers

The Indian farmers are cautious and usually tend to avoid taking risks. The provisions of the WTO stipulating reductions in exports subsidies on farm products will make Indian experts more competitive. The data on cost of cultivation, efficient agricultural practices and the availability of inputs will help to assess the strengths of indigenous products vis-à-vis imports. The availability of information of the adverse effects of the WTO regulations on any specific agricultural product will help farmers take corrective measures. In the emerging scenario, competitive advantage should be fully exploited to improve export potential. Opportunities for specialization may lead to better export potential. Similarly, forecasts on threats in terms of information related to cheaper imports and the macro-economic conditions of other countries are also required.

It is possible to lift geographical barriers by helping farmers come together online and facilitating the disposal of their produce at attractive prices. Online bidding can be introduced for various agricultural commodities. But this will require complicated IT systems, which should be supported by proper infrastructure and post-harvest technologies, storage, etc.

It is necessary to bring Indian farmers together for value additions to their agricultural output and to get them better returns from their produce. The opportunities for setting up such units, procedures related to exports, the quality norms to be adopted, packaging, and the like, must also be made available.

It is necessary to promote monitoring cells in all major institutions related to agriculture and allied activities to maintain data, provide periodic analytical reports and raise advance alerts.

Remote Sensing in Agriculture

The basic principle involved in remote sensing is that different objects reflect different amounts of energy in different wavelength ranges, based on different properties. A sensor is used to record such reflected energies from the surface of different objects. This recorded energy is then transmitted to the users, processed to form an image and then analysed to gather information about the target. In the context of frequent droughts and given the fact that weather forecasts are often wide off the mark, the use of remote sensing becomes imperative. The success of farm operations, from sowing to harvesting, depends upon correct forecasts. Moreover, it is also helpful in gathering vital information on pests, water and other natural resources depletion, etc, which can help planners and researchers effectively and strategically counter these. Also, it is possible to study continuously all kinds of renewable and non-renewable resources, such as water resources, forest cover, land use/land cover patterns, coastal resources, geology. Geomorphology, etc. Satellite remote sensing has great potential and is an efficient technology for the inventory and monitoring of natural resources.

However, remote sensing is not without constraints, these are:

- Non-availability of uniform spectral data.
- Data gaps due to cloud cover.
- High cost, time consuming.
- Low correlation yield for certain crops, like groundnut.

Expert Systems

Expert systems are currently considered the most commercially successful product of artificial intelligence research. The problems confronting farmers are complex. These include yield losses, soil erosion, diminishing market prices because of international competition, increasing chemical pesticide costs and pest resistance, and economic barriers hindering the adoption of farming strategies. Hence, farmers need to become expert managers of all aspects of their farming operations.

Expert systems are used to aid:

- Single-point decisions, such as the design of an irrigation system, selection of the most suitable crop variety etc.
- Sequence of tactical decisions throughout a production cycle, plant protection and nutrition decisions, etc.
- As a stand-alone advisory system for a specific case.

- As a delivery system for extension information etc.

Some agricultural expert systems in India and abroad are:

- Rice-crop doctors to diagnose pests and diseases for the rice crop and to suggest preventive and curative measures.
- Farm advisory systems to support agri-business management.
- Computer assisted agriculture, a system developed for farmers, agricultural officers and researchers with a facility to interact in the local language.
- Expert systems for cotton crop management for cotton growers.
- CALEX, a system developed by the University of California for pest control.
- A variety of systems for forecasting the needs of farming, cost-benefit analysis, water quality monitoring and crop rotation planning systems.

Constraints and Remedies for Effective Dissemination

Some major constraints, which are delaying the spread of the e-revolution in agriculture in rural India, are:

- Haphazard development, namely the duplication of efforts due to limited subjects. This can be overcome through a coordinating agency with advisory powers such as user interface, standards for setting up kiosks, etc.
- Non-availability of user-friendly technologies. This can be overcome by providing touch screen monitors and intuitive graphics-based presentations.
- Difficulty in providing advice in local languages through computers.
- The government's map restriction policies, which hinder the optimal utilization of data generated through remote sensing.
- Uncertain and poor quality power supply in rural India, which can be overcome by providing solar power packs for an uninterrupted power supply (UPS).
- Despite the fact that there are many facilities available,

connectivity in rural areas requires improvement.

The face of Indian agriculture can be transformed by a well-conceived deployment of IT. The key players involved in this process such as industry, government and educational institutions and research centres are required to male contributions in this endeavour. This initiative to develop necessary IT based agricultural services need to be developed immediately.

In the light of globalization it is necessary for the industry related to agriculture to review their present IT infrastructure with respect to marketing function and undertake measures to strengthen the same. No single institution or organization can succeed in the task of e-powering farmers and rural India. At the same time, scattered and half-hearted attempts cannot be successfully in meeting the objective. Industries with major stake in villages, should come together to provide the initial impetus.

The success of any IT based service to rural India hinges on evolving a proper revenue model for the dissemination points. The information kiosks can draw revenue from the industry by providing and disseminating required services. Once these dissemination points prove to be economically viable, the IT revolution in rural India will require no crusaders. However, it must be kept in mind that the challenges and rewards in this field are reciprocal and the greatest challenge is to keep pace with the ever changing and evolving technologies.

Future of Extensions

In view of the emerging competition in the fertilizer industry because of the increase in fertilizer consumption and crop productivity, the thrust areas of fertilizer promotion must be properly defined and tackled in order to get the desired results and improve the socio-economic conditions of farmers. Information is an important resource in agriculture. Farmers need information to improve their farming and extensions to provide them relevant and timely information. The main function of the extension is the transfer of information. These services can overcome obstacles in technology transfer and pave the way for agricultural development. Applications of communication technology in the transfer of agricultural information have two implications:

- The results provide the basis for adjusting strategies.
- Successful cases can become a model for replication.

Fertilizer marketing without promotion and extension is merely a selling activity. In a competitive market, service to the consumer/customer is most important. The availability of fertilizers has improved and farmers today have the option to select from various brands available in the market. Hence, extension education has become a dynamic concept in the fertilizer industry. Therefore, the current strategy of extension education is to keep farmers not only within the fold of balanced fertiliser use, but also to care for the maintenance of soil productivity. Extension services can guide farmers so that they have a proper mix of traditional and innovative farming practices- farmers no longer need to follow the legacy of their ancestors.

Structural and functional changes in agricultural extension are important to meet the needs and challenges of the new millennium. The globalization and liberalization era has made people re-think the role of public extension in developing countries. Extension workers/farmers can acquire knowledge as and when they have time and want to do so. For instance, a farmer may visit an extension office for pest control. He can identify the type of insect by consulting a database through photographs, consult the local entomologist and select suitable pesticides as a measure of pest control. The quality of extension material and process can be supplemented from time to time by multimedia, computer simulation and the like. Moreover, these extension systems can also make earnest efforts to organize farmers' groups and motivate and direct them so that they can demand appropriate and better extension services on a cost-sharing basis. The empowerment of small and marginal farmers through the cost-effective intensive use of electronic communication technologies is one of the most important possibilities.

Technologically, it is possible to develop suitable systems to cater to the information needs of the Indian farmer. User friendly systems, particularly in local languages, can generate interest in farmers. It is possible to create dedicated networks or harness the power of the Internet to make these services available to farmers in all parts of the country. But the task of creating application packages and databases to cater to the entire spectrum of Indian agriculture is a giant task.

Specialized Indian Institutions cater to various aspects of agriculture and can play a crucial role in designing the necessary applications, databases and services. Efforts should be made to promote a coordinating agency, which would have an advisory role and evolve a standard interface for users. This will call for urgent measures to introduce state-of-the-art technologies, such as remote sensing, geographical information systems, bio-engineering, satellite technology, and the like, to effectively monitor agricultural performance. This will not only help in planning, advising and monitoring the status of the crops, but also in responding quickly to crop stress conditions and natural calamities. But while developing these systems it must be borne in mind that the population that is targeted is not comfortable with the use of computers. And a major challenge is the dissemination of information.

4

Social Transformation through ICT

A. Thomas and M.A. Sudhir

"Progress of the villages is the true prosperity of a country".

The contemporary society is confronted with yet another factor of social change i.e. Information and Communication Technology (ICT). ICT leverages societal transformation, wealth generation and promotes integral well being of every citizen. It has brought about sweeping changes in all walks of life. In the recent past, the impact of ICT and the consequent changes in the society has become an area of study for the scholars as well as for the policy makers.

In the history of human civilization, technology has produced epochal changes in various societies. Technology can induce significant changes at different levels of the society. Today ICT, unlike many other technologies, is perceived and projected to have changed our environment to a large extent. It has not only altered our material environment but also our socio-cultural environment (Ogburn, 1965). Technologies have, slowly but steadily, entered the portals of rural India and the effects are visible to all the stakeholders.

ICT in Rural India

Rural India consists of 6,27,000 villages and 72.22 per cent of India's population lives in rural areas. The primary occupation of the people is agriculture. ICT finds a significant place in rural development, not only in terms of technological advancement but also in terms of reducing poverty. The incessant efforts towards computerization of rural development sector has facilitated the planning and monitoring processes, which are of direct significance to the public. ICT activity comprises of two components i.e. ICT Infrastructure and Application Software. 'Daily' is a dynamic web based application focussing on day-to-day activities and requirements of the Ministry of Rural Development (MRD) officials like, file monitoring, lodging of infrastructure related complaints etc. There are few other innovations that are only specific to Tamil Nadu i.e. Rural Bazar and PriaSoft.

Rural Bazar: Rural Bazar is a web-store that allows on-line marketing of the rural products. It can be configured to address different requirements and can act as a showcase for rural products. Its functionality can be extended to allow buying and selling with off-line/traditional mode of payments (bank draft or postal order). It can be further scaled to allow on-line payments along with off-line payment. Pilot implementation of Rural Bazar has been done at Kanyakumari and Kancheepuram districts of Tamil Nadu. Government of Tamil Nadu has provided a server to host Rural Bazar software and the data of products of afore stated district is being uploaded (http://ruralserver.nic.in/ruralbazar).

PriaSoft: Panchayati Raj Institution Accounting Software (PriaSoft) is a software package that addresses the monitoring of allocated funds, expenditure pattern, and local revenue generation and allows transmission of intended reports to various monitoring agencies. The software captures monthly data at grass-root level, subject to availability of computing infrastructure and transmits the consolidated data to the next higher formation through e-mail. PriaSoft empowers the administrators to monitor the fund receipt, availability, and expenditure at all levels of three-tier administrative set up of Panchayati Raj Institutions. It generates a number of reports and has an in-built decision support system to analyze any of the account heads. The software was demonstrated at Kancheepuram District of Tamil Nadu.

Occupational Diversification

One of the important features of the rural society in

general and Indian rural society in particular is its excessive dependence on agriculture. It is estimated that nearly 73 per cent of the Indian population is dependent on agriculture. The status of Indian agriculture has been worsening over the years due to excessive pressure on the land, increasing sub-division, vagaries of monsoon, lack of credit and support price for the produce. It is observed that even in green revolution areas where there has been spectacular increase in mechanisation and chemical inputs, there are severe problems related to infrastructural inputs like electricity, which can throw the best agricultural calculations out of gear. Naturally, the agricultural productivity has been dwindling and this falling rate of growth has been expelling people from agriculture. This negative trend called for certain acceleration in the growth of non-farm activity. Consequently, the non-farm activity did increase to a large extent and today it is estimated that nearly 25 per cent of the people are engaged in non-farm employment. In fact, concerted efforts have to be made to develop agricultural and allied activities to make agricultural not only viable but also remunerative.

Access to Market

The traditional markets were not farmer friendly on two grounds. First of all there were the middlemen who made huge profit out of the sweat of the farmer and secondly the farmer could not take advantage of the price fluctuations of the product. Naturally, the farmers were the losers on both the counts. Consequently, most of the small and marginal farmers, who constitute the major junk of the rural India, were forced to take credit from private financial institutions by pledging their crops and loose complete control over the price for their produce. Advent of ICT is able to dislodge middlemen and enable the farmers to negotiate with the probable buyer directly and wait for the best price. Moreover, when there are multiple markets adjacent to the farmers' village the farmer is able to know the prices at different markets and sells in that market that offers the best price.

Enhanced Knowledge Capital

For the rural people right to information is a fundamental right. If knowledge is power then ability to procure and process knowledge is an important entitlement. It is in this context that knowledge is treated at par with any other capital. As a capital along with financial capital, human capital, social capital and natural capital become an important livelihood resource. In other words, in itself and in combination with other capitals, it becomes a vital asset. Knowledge capital has four components:

- The extent to which poor have access to information institutions in the market, state and civil society.
- The ability of the poor to process and evaluate information.
- The extent to which the poor do not only consume, but also produce and share information within their community and network.
- The extent to which indigenous knowledge plays a role in the lives of the poor. ICT is the best available means of acquiring, storing and processing and transmitting this knowledge capital to the rural poor.

Conclusion

In the emerging knowledge society ICT has a pivotal role to play. India is moving ahead in her goal of emerging as knowledge superpower in the global scenario. In this dynamic situation rural development and progressive rural social change is central for the overall development of the country. ICT can achieve this critical need, provided that some of these concerns are handled proactively. Making ICT work for rural areas needs a total social re-engineering and social rejuvenation. Proper capacity building initiatives for the rural communities will certainly help the development agenda move ahead. This once again reminds that *survodaya* through *andyodaya* seems to be the way out for this country.

5

Interactive Multimedia and Farm Women

Ramasubramanian. M, Sujhi. G, Rathakrishnan. T, and Anandaraja. N

Gone are the days when women were recognized as subtle and submissive in all aspects. It is the "Era of Women Empowerment" where women are empowered not only with credit access but are enabled to be better entrepreneurs. The prosperity and growth of any developing nation like India highly depends on the status and development of women as they constitute half of its population. Women play multi-tasking such as looking after household responsibilities, caring and educating children, managing family and several other economic activities. This holds good for rural women who work with men in executing various field operations.

In India, women have a major place in agrarian scenario. Randhawa and Chandra (1993) stated that they play a significant and crucial role in various agricultural and allied activities besides fulfilling their responsibilities as home-makers. Sangwan *et al* (1996) reported that there is a distinct sphere of participation between males and females in farm and dairy sectors, the men were the planners and women implemented the activity. Their contribution is significant in the field of agriculture and allied activities as they are considered as backbone of food security.

Women and Agriculture Development

Rural women form a very significant part of the productive work force in the agriculture sector in India and most of the developing countries. Women are central to the selection, cultivation, preparation and harvest of food crops. Women farmers also possess a unique knowledge of dairying, animal husbandry and fish farming and handle most of the work associated with it. Verma (1992) indicated that women are responsible for 70 percent of actual farm work and constitute up to 60 percent of farming population. He also stated that role of women in agriculture is substantial and significant. In fact, historians believe that it was women who first started cultivation of crops and initiated the art of science and farming.

"Farm women continue to be the important stakeholders in farming activity of India. Presently, they constitute one third of the agricultural labour force and about 48 percent of self-employed farmers" stated Praveena *et al* (2005). Furthermore, management and involvement of Indian women in farming enterprise has been on rise in recent years especially in better endowed rural region. About 80 percent of all economically active women in the country are engaged in the agriculture sector. Also, about 48 percent of India's self-employed farmers are women. There are 75 million women engaged in animal husbandry as compared to 1.5 million men (www.digitalopportunity.org).

Contributions of farm women in agriculture cannot be ignored. The findings of Chaudhary and Sangram Singh (2003) support the above statement. They reported that women played considerable role in various agricultural activities such as pre-sowing, sowing and transplanting operations. Women's role was also reported in the preparation of compost and rising of vegetable nursery. Women's role was found to be maximum in hoeing and weeding practices on intercultural operations. Harvesting and post harvesting operations were reportedly more suited to the physical condition of women. Reaping of crops, storage of food grains, storage of seed and processing of grains were reported to be mostly done by women.

It is an acceptable fact that women possess talent which could be harnessed for the productive purpose. But it is most unfortunate that the role of women in agriculture has been neglected. A holistic development of women requires empowerment in all aspects involving them as managers, planners, scientists, and technical advisories. This is supported by Radha Singh 2005 as "A large part of women's contribution in agriculture is often not recognized and is also unpaid".

Gender-based Technology Dissemination in Agriculture

Agriculture is a world-wide critical strategic resource expected to double its production in 30 years to feed the growing needs of the population. Even now it continues to be the occupation and way of life for most of the Indian population. It is the lifeline of Indian economy. The sustainability of agricultural growth has been the focuses issue of the nation over the years. Transfer of technology generally aims at speedy diffusion of novel agricultural information from the place of invention (lab) to the place of adoption (land). For about five decades the information and communication tool has been conventional and only Home visit and Personal approach has been adopted by the extension personnel to disseminate the technical information to farmers. This method has been found time consuming and failed to disseminate the right technology to farmers at the right time who are spread across the country like India with diverse needs and interests.

So far all developmental programmes right from the Community Development Programmes to the recently launched Bharat Nirman were found to be men focused including the agricultural technology dissemination programmes. Despite multiple roles of women in agricultural operations and household chores, their work is generally underestimated and undervalued. This could be supported with the statement of Suman Singh and Puja Mathur (2005) as "the status of women in general is much lower than that of their male counter parts, largely because of customary male dominance in society, inherent shyness of farm women and lack of opportunities for education and training".

Paradigm Shift in Agricultural Extension

The role of agricultural extension is assuming greater importance in the context of globalization, privatization and liberalization and tremendous challenges of ensuring food and nutritional security to the Indian masses. A reorientation of existing strategies is essential to meet the new requirements of farming systems in the country and challenges in the areas of relevance, accountability and sustainability of agricultural extension. Ponnusamy and Jancy Gupta (2004) supports with their findings that "The paradigm shift is required to meet the challenges of the 21st century and to bring food, nutritional and environmental security and for serving all sections of the rural community including farmers, farm women and farm youth".

India should move towards an ever-green revolution, states an eminent agricultural scientist, Prof. M.S. Swaminathan, and this shoulders more responsibilities on the extension functionaries and the technology delivery systems to meet the diverse needs of farmers. Some of the paradigm changes in the agricultural extension are as follows:

- Group approach for agricultural technology dissemination.
- Involvement of community-based organizations for technology transfer.
- Formation of Self-Help Groups and extension through them.
- Rural location-specific knowledge centres for providing instant information to farmers.
- Need based problem-solving, skill based training for self-employment of rural youth.
- Harnessing the use of electronic media and digital revolution.

Potential ICT Tools for Transfer of Technology

The focus of technology development in agriculture in the last half of 20th century was to increase the production efficiency of the farm. For this the Indian farmers including the farm women should be updated with latest knowledge to compete for global marketing. They must have the right information at the right time which is possible through new information technologies resulted in many non-agricultural fields. Sharma *et al* (2005) indicated that the most commonly available IT tools for transfer of agricultural technology in Indian context are Radio, Television, Video, Telephone and mobile phone, Printed Media, Computer and Internet.

Information technology has rooted well as one of the tools of Transfer of Technology which are being used by both public and private sectors for the benefit of farmers. Village Knowledge Centres of MSSRF, E-Choupals of Indian Tobacco Company, Kisan Kendras of Rallies, Kisan Bharath Kendras of United Phosphorus Limited, Cyber Extension services of EID Parry's, Kisan Call Centres (KCC), AGRISNET AGMARKNET are some good examples to quote.

Self-help Groups (SHGs) and Technology Dissemination

Now-a-days group approach is employed to execute various rural development programmes. Usually women are mobilized into groups for undertaking mutually beneficiary social and economic activities. They are considered as powerful vehicles for development that function effectively through group dynamics. In fact, they also play an important role in technology dissemination. Malik et *al* (2004) stated that "Strategies for sustainable agriculture development and natural resource management should stress the participation and empowerment of farm women. The members of women SHG are from the most deprived section of the society, and are mostly associated with agriculture and allied activities for their sustenance".

"To give rural women visibility, they must be getting organized into Self Help Groups (SHGs). Group approach is a viable set up to empower women economically, socially and technologically for improved life", reported Suman Singh and Puja Mathur (2005). Role of SHGs is emerging as a promising tool in all aspects. The SHGs are created to enable the members to reap economic benefits of mutual help, solidarity and joint responsibility towards self and sustainable development".

Group Approach to ICT Tools

Trends of development in ICT indicate that it following the usual paths towards inequality and exclusion, putting women and girls at a disadvantage with regard to securing better livelihoods and other rights. Information technology is making a significant contribution in the empowerment of women. The women SHGs and their members are using ICT approaches in both farm and off-farm activities. The government institutions are also engaged in the development and popularization of gender friendly tools and equipments. With this key idea in mind, development of user friendly multimedia for enabling farm women has been initiated in research institutions.

Role of Multimedia

Any communication should be very effective in conveying the message such that it is properly understood and location specific, interactive, self-dependent and fast in delivery what is intended to be delivered. The traditional methods of communication lack one or the other above features and hence could not provide holistic information. Computer multimedia will offer learners more complete and individual control over their learning. The main reason behind recommending the use of computer based multimedia system for farmers to facilitate interactivity and better understanding between individual learners and the subject matter. Multimedia tools are ideally suited to demonstrate complex and dynamic process that cannot be explained easily with conventional media and methods.

Multimedia is a modern electronic gadget with a combination of text, graphics, audio and video allowing the end user to control the delivered elements. Aeron (1998) indicated that, if you are stimulated with audio, you will gain about 20 percent retention rate that of audio visual is up to 30 percent and in an interactive multi media presentation, where you are really involved, the retention is as high as 60 percent.

Swanson *et al* (1990) reported that person-to-person communication has traditionally been the most important form of information transfer. Print media as well as radio and television were of supplementary nature because they frequently lacked target group or location specificity and information was not up-to-date.

Revolutionary changes in communication technology have dramatically increased the speed and quality of information transfer and changed the role of extension workers in industrialized countries. Electronic communication systems may in part replace personal visits, and one of the major tasks of any agent will be of link her or his clients with other suppliers of information.

Multimedia is a Combination of Content Forms

Multimedia presentations may be viewed in person on stage, projected, transmitted or played locally with a media player. A broadcast may be a live or recorded multimedia presentation. Broadcasts and recordings can be either analog or digital electronic media technology. Digital online multimedia may be downloaded or streamed. Streaming multimedia may be live or on-demand. Enhanced levels of interactivity are made possible by combining multiple forms of media content.

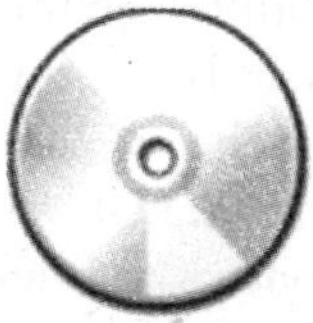

Local Recorded

Online Streaming

The speedy diffusion of novel agricultural technologies could only be achieved through active involvement and participation of farm women with the help of IMCD. IMCD goes a step further in providing information to the individuals / farm women round the clock in local language using audio and video clipping. IMCD is increasingly attracting the attention of farmers including farm women, extension functionaries and educationists as a way of much-needed speedy ToT in agriculture. It is fast becoming the most successful tool for ToT and imparting education in agriculture. In fact, it is a motivational cum inspirational tool playing a significant role in sweeping changes in socio-economic development in the era of globalization. In the light of the above factors, a research work on developing a farmer friendly IMCD as an effective tool of technology transfer for farm women has been carried out.

Various initiatives are in vogue to enable farm women not only in ICT but also in certain other key areas. The Common Minimum Programme provides that the government should ensure that at least one third of all funds flowing into panchayats should be earmarked for programmes for the development of women. The National Commission on Farmers set up by the government also has its terms of reference recommending measures for credit, knowledge, skill, technological and marketing empowerment of women, taking into consideration the increasing feminization of agriculture.

A National Gender Resource Centre in Agriculture (NGRCA) has been set up in the Ministry of Agriculture as a focal point for the convergence of all issues related to women in agriculture. A similar step has been initiated by the Department of Agricultural Extension, Agricultural College and Research Institute, Tamil Nadu Agricultural University which is operating a scheme entitled "Diffusion of Farm technologies to Farm Women through user-friendly Interactive Multimedia Compact Disc". The scheme has envisaged selecting two villages under different eco-systems followed with identifying the felt technological needs of the farmers through Participatory Rural Appraisal (PRA) and Focused Group Discussions (FDGs). The identified and prioritized technological needs would then be developed into a user-friendly interactive IMCD. Using the IMCD, latest and updated agricultural technologies will be transferred to the formed women user groups thereby increasing their knowledge and skill. In this regard, around four to five IMCDs have been developed in the scheme namely System of Rice Intensification for Paddy, Eriophid mite control in Coconut, Vermicompost production etc.

IMCD on Vermicompost Production

Vermicompost production, widely recognized as an important component of organic farming, has been identified as the right technology for dissemination. Transferring this technology to farm women would not only enable and empower them as entrepreneurs but also enriches soil fertility increasing the agricultural production of the country. It also protects and safeguards the environment from all hazardous pollution. IMCD on Vermicompost Production is the recently developed IMCD which has contents on various techniques of Vermicompost production, Vermiculture, Inputs required, Cost of production, Price details and Income benefits in a more comprehensive, interactive and user-friendly manner. This has been developed as it is felt the most viable entrepreneurial activity for farm women. It has also been developed with the main motive of overcoming language barriers, illiteracy constraint as it is produced in local language with graphics, video images and animation so that all can easily access it.

Conclusion

Women who are being considered as backbone of food security have to be imparted education and knowledge on innovative agricultural technologies for holistic empowerment. In addition, they should be involved in transfer of technology for faster diffusion. This could be true with the words of Jawaharlal Nehru "When women move forward, the family moves, the village moves and the nation moves". For transfer of technology high retention rate is highly required and IMCD offers as high as 60 percent retention rate in imparting knowledge. A famous quote reminds that "Change is the only constant in the world" which means globe is dynamic and keeps changing more often. Today we are in the Information and Communication Technology (ICT) world highly depending on computer and allied modern electronic gadgets for executing our day to day activities. Almost all fields are being influenced by the interventions of ICT and agriculture is no way an exception to that. Hence, Interactive Multimedia Compact Disc has been recently recognized as one of the ToT tool to educate and impart knowledge on innovative agricultural technologies to farm women who are considered as prospective agricultural extension clientele. IMCD is highly pragmatic and observed to be a new recipe for empowering farm women.

References

Aeron, R.K. 1998. "*What is Multimedia, Multimedia Production Concepts? Summer School on Computer Multimedia Application in Agriculture and Allied Sciences*". College of Technology and Agricultural Engineering, RAU, Udaipur, India.

Anandaraja, N. 2002. "*Developing Farmer Friendly Interactive Multimedia Compact Disc and Testing its Effectiveness for Transfer of Farm Technology*". Unpub. Ph.D., Thesis. AC&RI, TNAU, Coimbatore.

Anandaraja, N., Rathakrishnan, T., Durai Singh., and M. Ramasubramanian. 2006. "*Interactive Multimedia Compact Disc: An Innovative Approach to Disseminate Farm Technology*". The Fifth International Conference of the Asian Federation for Information Technology in Agriculture. Nov 9-11, 2006. Bangalore. India. pp.150-156.

Chaudhary, H., and Sangram Singh. 2003. "*Farm Women in Agriculture Operations*". Agricultural Extension Review. 15(1): pp. 21-23.

http://en.wikipedia.org/wiki/Multimedia

Jasbir Singh Manhas., Meena, B.S., Charak, A.S., and V.P. Sharma. 2005. "*Potential IT tools for Transfer of Technology*". Agricultural Extension Review. 17(2): 10-12

Malik, B.S., Das, B.C., and V.K. Yadav. 2004. "*Women Self Help Group – Disseminate Agri Information*". Agricultural Extension Review. 16(1): 10-14.

Ponnusamy, K., and Jancy Gupta. 2004. "*Agricultural Extension – New Paradigms*". Agricultural Extension Review. 16(6): 3-6

Radha Singh, 2005. "*Gender issues are central to the development dialogue*".www.digitalopportunity.org.

Randhawa, A., and S. Chandra. 1993. "*Changing role of Home Scientists in Transferring from Technologies to Farm Women in Agriculture*". Paper presented at National Seminar on Women in Agriculture. Developmental Issues at NAARM, Hyderabad. December, pp. 28-30.

Sangwan, V., Munjal, S., and P.K. Punia. 1990. "*Participation of Farm Women in Farm Activities*". Indian Journal of Extension Education. 26(1&2): 112.

Suman Singh and Puja Mathur. 2005. "*Self Help Group – A Successful Approach for Women Empowerment*". Agricultural Extension Review. 17(6): 27-30.

Swanson, B.E., Farner, B.J., and Bahal, R. 1990. "*The current status of agricultural extension worldwide*". In FAO, Report of the global consultation on agricultural extension, Rome: FAO, pp. 43-76.

Underdahl, Brain. 2001. "*Teach yourself Microsoft Office 2000*". IDG Books India (P) Ltd, New Delhi.

Verma, S.K. 1992. "*Women in Agriculture: A socio-economic analysis*" Concept Publishing Company. New Delhi.

6

Digitalization of Indigenous Technical Knowledge (ITK) for Farming Community

Suganthi N.S, Poonthenmozhi. S, Rathakrishnan. T, Anandaraja. N and Ramasubramanian. M

Indigenous Traditional Knowledge (ITK) is a valuable resource for the development of rural community under certain circumstances; it can be equal to or even be superior to know-how introduced by formal research system. Development efforts should therefore consider ITK and use it to the best of advantage. Although more and more development professionals have come to realize the potential of ITK, it remains a neglected resource. A key reason for this is lack of guidelines for documentation and preservation of ITK. Documentation and preservation of ITK through digital mode is one of the best methods for exploration and better preservation.

What is ITK?

In agriculture, for instance, this knowledge is shown in the development and adaptation of plants and crops to different ecological conditions (soils, rainfall, temperature, altitude, etc). Traditional knowledge is people's awareness and understanding of this and other information, which is passed on from one generation to the next, usually by word of mouth or example within a specified group of people. Indigenous knowledge is often used interchangeably with traditional knowledge.

Importance of ITK

In India, still considerable portions of the farming community have not yet adopted the modern farming techniques. However, some of the improved technologies do not find place equally in all the cases because of local variations in multi factor scenario. Similarly it requires costly inputs and complex to understand and practice by the small farmers. Importance of the study of indigenous knowledge system was reported by several authors. Scones (1989) pointed out that understanding of farmers' knowledge allowed a framework of reference for posing technical and scientific questions in research and also provided the basis for evolving technological options.

Warren (1991) reported that it was important to recognize indigenous knowledge and organizational capacity as national resources as they facilitated national development. Farmers had been actively involved in identifying and prioritizing agricultural problems and extending information. He added that farmers continued with their own research and extension agenda outside the public and private sector institutes' setup within a nation to provide the same functions.

Fujisaka (1992) pointed out that the farmers and researchers needed to contribute to the development of improved and adoptable upland rice. He remarked that farmers could contribute to selection criteria about which researchers currently know little, such as deterioration of taste over time, hardening of cooked left over, or expansion when cooked. On the other hand, researchers would continue to investigate such things as nutrient uptake dynamics, mechanics of drought resistances and different strategies for seeking blast resistance. Dubey (1993) mentioned that the traditional methods and techniques were fruitful and beneficial in the sense that they were not generating ecological imbalance and help the peasants to sustain with their meagre resources owing to population explosion and related problems. Rajasekaran (1993) stated that the study of indigenous knowledge system helped in understanding the perspectives of local people, bridging the communication gap between outsiders and insiders, recognizing the accomplishments of local people, helping outsiders familiarize with the local conditions and increasing the participation of farmers and their organizations in integrating, utilizing and disseminating what already exists.

Tabor (1993) pointed out that indigenous systems had an advantage in that they were widely known by the people of the region and could easily improve communication between farmers, extension agents and scientists. Quiroz (1996) reported that there was strong empirical evidence that farmers are actually quite active experimenters. The majority of small-scale farmers in developing countries live in conditions of high variability, uncertainty and complex interactions. For them indigenous innovations, which are generated mainly through experimentations, represent one of the key elements in their survival. The author further added that agricultural knowledge of farmers could play an important role in the conservation of bio-diversity and this indigenous knowledge system could facilitate development.

Documentation of ITK

ITK is often transmitted by word of mouth rather than written document. This makes it vulnerable to rapid change-especially when people are displaced or killed in famine or war, or when younger generations acquire values and lifestyles different from their ancestors. Some ITK is lost naturally as techniques and tools are modified or fall out of use. During the last decades, however, development processes and population changes have accelerated this loss, endangering the survival of ITK. Hence ITK has to be documented and preserved for serving the farming community and saving farming.

What can we do to preserve this endangered ITK?

- Raise awareness about the value of ITK for development.
- Help communities conserve their ITK.
- Record and use ITK in applied development projects.

- Document ITK and make the information available to people working in development.
- Make ITK available to the communities from which it was obtained.
- Observe Intellectual Property Rights when recording ITK.

Methods of Recording Indigenous Knowledge System

Various methods have been suggested by different authors for effective documentation of farmers' indigenous practices. Gupta (1990) reported that collection of indigenous knowledge could be attempted by many different ways like farmers' participatory research and Rapid Rural Appraisal (RRA), which can raise awareness of rural poverty and associated problems. He also stated that Manually Discriminated Analysis is a method, which draws upon farmers' own innovative genius. The basic assumption is that by comparing and contrasting the farmers' own practices a hypothesis can be developed.

Mundy and Compton (1991) stated that the sources of indigenous information could be categorized as:

- Indigenous experts (such as farmers particularly skilled in raising goats).
- Indigenous professionals (such as healers and irrigation specialists).
- Innovators (people who experiment with and develop new techniques).
- Intermediaries (those who pass on messages such as 'town criers' and messengers).
- Recipient disseminators (all those who receive information, modify it and pass it on).

Rajasekaran (1993) recorded indigenous knowledge systems using farmers' participatory methods such as participant observations and unstructured interactions. According to Dubey et al. (1993) several techniques like case study method, oral history method, and key informant method, making diagrams, case histories, critical incidents, preference ranking and inventory of farmers' indicators could be used for eliciting and documenting the knowledge from local people. Mane and Sutaria (1993) used the methods like dialogues, field observations and joint interpretations to arrive at conclusions to document the indigenous practices.

Massaquoi (1993) reported that the method of obtaining information on the indigenous technology would depend on the particular technology and could include two or more of the following:

- Observation of the technology in operation,
- Interviews with users, and
- Controlled trials of the traditional technology.

Singh (1993) used individual and group interviews, participant observations and agro-eco system analysis for recording indigenous knowledge system. Vivekanandan (1993) mentioned that conducting village level workshops, group discussions with farmers, artisans, publication of news letters in local language for exclusive communication of traditional farm technologies and travelling to interior regions were some of the effective means in identification, documentation and dissemination of traditional technology and also a means of getting feed back from the people. Arulraj and Vasanthakumar (1996) utilized participatory Technology Development (PTD) approach to identify indigenous knowledge and to integrate the knowledge in technology development process.

Digital Documentation of ITK

We now hereby discuss documentation of ITK results and transfer of knowledge across the generations from one community to another community.

Advantages of Digital Information

- The digital information is relatively less expensive than the printed version.
- High quality.
- Attractive functions, such as selective dissemination of information (SDI), PDF (Portable Document Format) and

document delivery.

Disadvantages

- Need of proper training to information professionals.
- High technical requirements (electricity, communication, net connection).
- Hands on experience and training.

Major Threats to ITK

1. Loss of the indigenous people's territorial base through the destruction of forest resources and biodiversity.
2. Introduction of the modern technologies in the field of agriculture and others.
3. ITK is increasingly endangered by mis-appropriation of this knowledge by outside researchers. So the valuable treasure of our society, the ITK should be documented and preserved in easily available and retrievable format. In order to achieve this, ITK should be documented by using modern electronic gadgets (video camera and digital camera) i.e. writing a novel things with old pen filled with new ink.
4. Previously ITK has documented through written documents, drawings, audio recordings.
5. Written documents, drawings can easily be damaged as the day passes by. Another drawback is illiterate people can't utilize this document for their reference and future use.

Audio recordings: Regarding audio recording only voice of the respondent can be recorded and reproduced to other people in the community, for their future use and adoption. But no visual image/picture can be seen.

The digital documentation replaces all these disadvantages, because digital documentation will give the entire process of the technology with audio and video. So digitalization of ITK is more and more advantageous over the conventional type of documentation and preservation.

Documentation through Digital (Photo) Camera

Images are important to keep the collective memory of the

agricultural traditional practices and keep these alive, and photography is increasingly contributing to this. This is especially relevant today, when local knowledge is rapidly being lost as a result of migration and the undervaluing of local traditions and culture.

Advantages of Digital Photography

- Digital cameras are very simple to use and it is quite easy to get a high quality photo. Different modes are available to get clear picture, e.g. L, M_1, M_2, M_3 etc.
- Films are not needed, nor are laboratories necessary for developing the photos. Instead of films memory card with various memory capacities is used to capture the images /pictures.
- Images are visible almost immediately with the help of LCD display and may be duplicated at no extra cost.
- Unwanted photos can be deleted then and there.

Documentation by Using Digital Handy Camera

- Video camera allows documenting the ITK with voice and visuals i.e. audio and video. If a picture speaks a thousand words, a video speaks a million. Video documents can also be easily copied into CD ROM and can be disseminated through internet. In this way video documentation can being local experience and knowledge into global network.
- Previously cassettes are used to capture the video, nowadays mini DVD (with various capacities) can be used to record/capture the video and captured video can be seen immediately and conversion for replication and dissemination also easy.
- After completion of the visual documentation, the captured images can be easily transferred from mini DVD to system by using Nero Express software. After that DVD can be formatted for future documentation.
- While documenting the technology by using digital handy video camera, we can adjust the video capturing by viewing through LCD.
- Through visual documentation entire process of the

technology can be reproduced to the community for their better understanding, because seeing is better than hearing.

Softwares used for Editing the Visual Documentation

1. Adobe premier.
2. U-Lead.
3. Pinnacle studio.
4. VCD cutter.
5. Arch Photo studio.

Preservation of Documented ITK

Many initiatives have been taken to consider the preservation and protection of traditional knowledge base (TK). The thrust of most of the national an international initiative as TK is to develop a system that documents and preserves TK created in the past and contribution to promotion and dissemination of innovations.

Such a system might provide appropriate information for continuing creativity and innovation is most of fields of including traditional medicinal and agricultural practices.

- An important step is to establish digital data base to enable quick retrieval of information of TK for a variety of uses. Information on TK is also significant for researchers and scientists. The database may also meet the requirement of R& D scientists.
- Traditional knowledge Digital Library (TKDL): Traditional knowledge Digital Library, a collective effort between NISCAIR (National Institute of Science Communication And Information Resources) and Indian system of Medicine and Homeopathy (ISM&H; Ministry of Health and Family welfare), was launched on March 27, 2002. TKDL is a storehouse of current traditional knowledge about plants and the ways and means to threat the disease with traditional medicine.
- In India, a Traditional Knowledge Digital Library (TKDL) has been set up to record details of medicinal plants (currently 4,500), in an easily searchable database. This

allows "inventors" to make searches of the database to check if they can patent their product.

- Traditional Knowledge Data Library (TKDL) will make available the widely scattered and distributed references on our traditional system in a retrievable form.
- TKDL acts a bridge between traditional and modern system of agriculture.
- Availability of this knowledge in a retrievable form in many languages will give major impetus to modern researcher.
- Sustained efforts on modernization of our traditional system will create increased awareness at various levels (state and national level).
- Bio piracy can be avoided through patenting. Electronic databases and digital libraries are gaining popularity in several government-initiated projects for documenting traditional knowledge.

Technology Information Facilitation Programme (TIFP)

TIFP has the broad objectives of generating endogenous capacities for the development and utilization of digital information resources to facilitate accelerated science and technology (S&T) research. In an endeavour to ensure that information and communication technologies (ICTs) are fruitfully used in all the sectors of development and facilitate collaborative research among industries and institutions, the programme outlines as below.

The specific objectives of the programme are designed to:

- Develop appropriated endogenous information capacities to support R&D activities.
- Support the production of local content and to promote capturing of indigenous knowledge base.
- Promote information and knowledge networking at local regional and national levels to facilitate flow and sharing of information resources.
- Support education, training and R&D in digital content

development and utilization.

- Promote national and international cooperation in related areas.

Department of Science and Industrial Research (DSIR) has given the project entitled "Documentation and Preservation of Agricultural Traditional Knowledge through Farmer Participatory Approach" for digital documentation of ITK to Department of Agricultural Extension and Rural Sociology, AC & RI, Madurai.

Objectives of the Project are:

- To identify the traditional knowledge through farmer participatory mode in agriculture and allied field.
- To document the traditional practices with the help of electronic gadgets like digital, handy camera and print media.
- To ascertain the scientific rationale behind the identified traditional knowledge.
- To explore the possibilities of patenting and obtaining intellectual property rights.
- To release and launch of multimedia websites, multimedia compact discs both in local and international linguistics for wide spread dissemination.

In Tamil Nadu more than 65 percent of the population is engaged in agriculture and state economy largely depends on the agriculture sector. Tamil Nadu Agricultural University (TNAU) is one of the reputed Universities in our country successfully running with the objectives of research, extension and education. Every year, the University is releasing a number of need based crop varieties and management innovations for the farming community and focusing on the research and development works on all the agricultural related aspects. Besides, by using the latest technologies developed in TNAU, the farmers are also following traditional Indigenous Technical Knowledge (ITK) system for agriculture and allied fields.

The importance of indigenous technology and practice to

sustainability is being brought through pooling of traditional knowledge, short listing and evaluating them in the context of modern scientific and technological environment and harnessing it for sustainable agriculture growth. For achieving sustainability in agriculture, ITK is essential and it should be identified, classified and documented for refining, supported with scientific reasons and mixing with scientific technology through various modes for its immediate acceptance and adoption of the technology in the user system without any time lag, since time lag is the prime factor in the diffusion and adoption process of any technology.

Use of indigenous knowledge in the development projects generally facilitates the participation of the beneficiaries. A blend of indigenous knowledge and modern technology may be most appropriate for sustainable development. Hence, this study will pave way for sustainable agriculture development.

Table 6.1: Total Number of ITK Digitally Documented So Far

Particulars (District)	Trichy	Namakkal	Nagapattinam	Erode
1	**2**	**3**	**4**	**5**
Traditional varieties	2	9	7	5
Cultural practices	22	-	6	5
Implements	6	-	2	10
Storage/Post harvest practices	20	-	3	6
Medicinal plants	24	2	1	6
Livestock management	14	-	5	3
Fishery	-	-	2	-
Total	88	11	26	35

Contd...

Particulars (District)	Krishnagiri	Sivagangai	Kanyakumari
1	**6**	**7**	**8**
Traditional varieties	7	-	31
Cultural practices		11	22
Implements	9	-	14
Storage/Post harvest practices	7	1	14
Medicinal plants	11	-	30
Livestock management	11	66	1
Fishery	-	-	15
Total	45	78	127

Under this scheme, 410 traditional practices have been documented in six districts of Tamil Nadu, in a digital mode by using handy camera and digital camera. A series of different micro-level studies needs to be conducted in different agro climatic zones to have an exhaustive list of the same. Further, their scientific rationality and economic feasibility should be determined. The practices which are found to be rationale and feasible should be tested at research stations. Scientifically sound and economically viable traditional practices can be recommended for adoption in other areas having similar situations. Study of traditional knowledge and farmers' rationale can also encourage 'participatory research' and 'participatory technology development' process. Technology blending program should be started where traditional and modern technology could be welded together in order to generate appropriate technology.

Conclusion

After completion of the documentation, scientifically valid ITKs will be used to promote awareness, exchange within the same community and in other communities. ITK has become economically important; many initiatives have been made to conserve the ITK. This may also facilitate the deserving, to reap the benefits of the ITK for commercial gains. The documentation and setting up of digital data bases of ITK are of prime importance for this purpose. So ITK can be documented, preserved and stored in a digital mode. Intellectual Property Rights will be applied for scientifically validated practices for welfare of the farming community. This will pave way to safeguard farming community from biopiracy.

References

Anne Pipenstock,OrlandoArratia and Luis Carlos Aguilar.2006."*New technologies support farmers documentation*". LEISA INDIA, Vol8 (1). 31-32.

Arulraj, S. and Vasanthakumar, J. (1996) "*Participatory Technology Development: A case study*". Discussion paper 9611. Coimbatore: Sugarcane Breeding Institute.

Chris Lunch. (2006). "*Participatory video as a documentation tool*". LEISA INDIA, Vol 8(1). 33-34.

Dubey, B. (1993) "*Agricultural Paradigm in Indian Tradition. Paper presented at the National Seminar on Indigenous Technologies for Sustainable Agriculture*", New Delhi. March 23-25.

Fujisaka, S. (1992) "*Where there has been no Green Revolution Farmer's Upland Rice and Related Technology in Mindan, Philippines*". Paper presented at the International Symposium on Indigenous Knowledge and Sustainable Development.Silang. September 21-25.

Gupta, A.K.(1990) "*Documenting Indigenous Farmers' Practices*". ILEIA News Letter.6 (2):29.

Gupta. V.K.(2001)."*Report of the task force on traditional knowledge digital library*": Gist. Journal of Intellectual property rights. Vol. 6:121-133.

Mane, P.M. and Sutaria, N. Jivaraj (1993) "*Study and Documentation of Indigenous Knowledge and Traditional Agricultural Practices of the Tribal Farmers*". Paper presented at National Seminar on Indigenous Technologies for Sustainable Agriculture. March 23-25.

Masaaquoi, J.G.M.(1993) "*Indigenous Technology for Off Farm Rural Activities. Indigenous Knowledge and Development Monitor*".1 (3): 2-3.

Meenakshi prajneshu and Gupta. (2001). "*Traditional knowledge data base*": IPR and opportunities for R&D, Journal of Intellectual property rights. No. Vol. (6): 449-458.

Mundy, P.and Compton, L.(1991) "*Indigenous Communication and Indigenous Knowledge*". Development Communication Report. 74 (3): 1-3.

Quiroz, C. (1996) "*Local Knowledge Systems Contribute to Sustainable Development. Indigenous Knowledge and Development Monitor*": 4(1): 3-5.

Rajasekaran, B.(1993) "*A Frame Work for Incorporating Indigenous Knowledge Systems into Agricultural Research, Extension and NGOs for Sustainable Agricultural Development Studies in Technology and Social Change*" No.21 Ames, lowa: Technology and Social Change Programme, lowa State University.

Scones, I. (1989) "*Patch Use by Cattle in Dry land Zimbabwe: Farmer Knowledge and Ecological Theory*". ODI Pastoral Development Network Paper No, 28.London: Overseas Development Institute.

Singh And Singh.(1993). "*Neem The wonder tree for all*". Paper presented at the National seminar on Indigenous technologies for sustainable Agriculture, New Delhi, March 23-25.

Tabor, T. A.(1993). "*Soil survey and indigenous soil classification. Indigenous Knowledge and development monitor*". 1(1):28

Vivekanandan, P.(1993) "*Approaches in Documenting Traditional Technologies: Process and Outcome in Tamil Nadu*". Paper presented at the National Seminar on Indigenous Technologies for Sustainable Agriculture, New Delhi. March 23-25.

Warenn, D.M.(1991) "*The Role of Indigenous Knowledge in Facilitating the Agricultural Extension Process*". Paper presented at the International Workshop on Agricultural Knowledge Systems and the Role of Extension, Germany. May 21-24.

7

Village Knowledge Centres (VKCs) and Agricultural Productivity

Samwel Kakuko Lopoyetum, K. Rakes Sarma and Rangamlian Aimol

Introduction

Information and communication technology is a vital component of knowledge based agricultural productivity in India. It is essential to ensure that the benefits of ICT reach all the socio-economic stratum of the rural population in India. Village Knowledge Centres (VKCs) can provide crucial services to the rural farmers by using advanced and modern information and communication technology. Hence, setting up of VKCs to provide interactive sessions and also to the remote villages will be a boon to rural development (rural economy). New technologies supported by appropriate services and public policies have led to agricultural revolution, improvement in production and productivity of rice, maize, soybean, potato, fruits and vegetables, green gram, mustard plants, bananas, sugarcanes, groundnuts, sweet potatoes and many other crops as well as farm animals (dairy animals) in many developing countries. A hunger free country can be possible, if we pay adequate attention to improve food availability through higher agricultural productivity and use of better ecologically sustainable methods of production. Farming has become knowledge intensive by using Geographical Information System (GIS), Global Positioning System (GPS), latest information, computer technologies and satellite imagery to target specific areas such as application of fertilizers, pesticides, water use, planning, administration, farm management, etc.

The ICT revolution allows for rapid growth in systematic assimilation and dissemination of relevant and timely information and access to universe of knowledge and communication through low cost electronic materials and equipments. The software industry is providing new tools to increase productivity and create new opportunities for understanding complex agro-industrial system, intensive agriculture etc. Knowledge Connectivity to villages through VKCs will act as a frontline delivery system to provide essential data and information required for the targeted population such as farmers, dairy farmers, fishermen, craftsmen, rural traders, businessmen, unemployed youth and over all rural economy. Application of technology for improving agricultural productivity is necessary in rural areas particularly in the changing economic environment.

The typical knowledge required in agricultural production extends to soil science and management, seeds, water management, pre-harvesting practices, agronomic practices and post-harvesting management, productivity increase, banking (loans), financial assistance, technical assistance to farmers, meteorological forecasting for disaster warning etc. VKC can use advanced technology and can promote technology for sustainable development and create knowledge society. Agriculture in India is dominated by resource and information–poor farmers hence can get boost through VKCs and a better agricultural extension approach. The setting up of information centres/kiosks at village level or micro-level with facilities using computers and access to Internet can offer reliable and upgraded multipurpose input services. It must ensure an effective dissemination of the latest and need based information to rural farmers. It is heartening that M. S. Swaminathan Research Foundation (MSSRF) had initiated VKC in Pondichery in 1988 to provide rural families with access to information, fostering agricultural and rural development through the use of ICT.

The content of dissemination is prepared using indigenous

knowledge with generic information including those propounded by the Subject Matter Specialist (SMS) and use of Multimedia Interactive Systems. The changing environment is moving towards an era of knowledge intensive agriculture. This can be effectively done through Rural Knowledge Centres (RKCs) across the country. More than 70 percent of India is rural and depends mainly on agriculture and allied activities but ICT led knowledge in the rural economy is absent or dismal. Therefore, they must be brought under VKCs.

More than 600,000 villages can benefit from advanced ICT technologies and it can make every village a knowledge centre to trigger knowledge economy revolution in the rural areas. A sustained increase in agricultural productivity (increase in yield per acre, increase in yield per cow, increase in water use per unit and effective water usage) is very crucial to the growth of rural economy. ICT is one factor, which can contribute to agricultural productivity. Such variables as reliable irrigation, quality seeds, fertilizers, organic fertilizers, inorganic fertilizers, use of pump set, electricity (these are all Green Revolution Strategies) have assumed greater significance in the Indian agriculture in the changing economic environment. There is an imperative need for increasing the agricultural productivity (non-foodgrains and foodgrains productivity). The potentiality of attaining higher productivity for food crops and non-food crops is possible through VKCs, even though there are numerous causes for low agricultural productivity in India such as illiteracy, ignorance, superstition, conservativeness, caste system, primitive way of cultivation, poor techniques of production and poor technology.

Agriculture development is central to India's economic development. The total socio-economic dimensions of development are basically gained through agriculture. Due to rapid population growth, we are not able to produce sufficient food for nutritional security for each and every one. Malnutrition incidents have had occurrence in Maharashtra and Orissa recently. Agriculture is the backbone of many developing countries such as India and Kenya. The livelihood of almost more than 65 percent of Indian population is dependent on agriculture and allied sectors. Twenty eight percent of National Gross Domestic Product (GDP) and 19 percent of the country's export income comes from agricultural sector. Understanding the role of agriculture on the country's economy, our first Prime Minister, Jawahar Lal Nehru opined, "Everything else can wait but not agriculture". Hence, "everything can wait but not ICT". ICT application in enhancing agricultural productivity in the twenty first century is therefore imperative. From its inception onwards, Indian agriculture has been receiving the top most priority by making allocations of the large planned investment on different heads of agriculture. The monitoring and evaluation of such investment should receive the attention of all so as to achieve a significant benefit over expenditure and for food and nutritional security of country's burgeoning population.

We should look back to understand how the Indian agriculture in terms of growth, production, productivity and overall performance tends to gain through ICT and accordingly govern its direction towards 21st century and at the same time provide basis of its past performance. Even if we had achieved a lot of productivity improvement in our agricultural sector after the post-Independence era, it is sad to say that during the pre-Independence era, the British Rulers had swallowed all natural resources (both in rural and urban areas), including agricultural produce/commodities. The Export Processing Zones have taken place in India due to the predominance of agriculture, agro-based industries and allied activities. Hence, there is an imperative need for the expansion of ICT and VKC's in all the villages in rural areas to promote EPZs / SEZs in the changing economic environment.

Food and Nutritional Security through VKCs

The Indian planners, right from the beginning, realized the need to attain self-sufficiency in food grains as one of the important goals of socio-economic planning. The Government realized that food surplus countries used food surplus as a weapon to force several countries (particularly, those in developing countries) to submit to their dictates. All these Governments have used ICT to popularize for higher agricultural productivity in their countries. Prime Minister Jawaharlal Nehru realized that it was with great difficulty that India was able to avoid the political strings attached with food aid, but it did hurt national pride. In one of his broadcasts to the nation, Nehru stated very candidly: "We have sought help from abroad... and we shall continue to do so under pressure or necessity, but the conviction is growth upon me more forcefully than ever how dangerous it is for us to depend for this prime necessity of life on foreign countries. It is only when we obtain self-sufficiency in food that we can progress and develop ourselves. Otherwise, there is the continuous pressure of circumstances there is trouble and misery and there is sometimes shame and humiliation". In order to avoid this problem, rural farmers must be given the benefits of ICT to increase and enhance their agricultural productivity and standard of living.

The Government of India under Prime Minister Indira Gandhi went in for seed-water-fertilizer policy popularly known as the Green Revolution. This policy ushered in a revolution in food production in India and dispensed with food grains imports altogether. India succeeded in raising her food grains production from 51 million tones in 1950-51 to 212 million tones in 2001-2002. India had achieved self-sufficiency in food grains by the year 1976 and since then, Indian imports of cereals have remained negligible. Even Gilbert Etienne while appreciating India's achievements on the food front states: "Despite the gloomy and perfectly ill-founded forecasts of 'prophecies' about India's future disasters, which were fashionable in the 1960s the country is no longer exposed to real famines".

Production of food grains front reveals the following status

- Between 1950-51 and 2008-09, food grains production has increased from 51 million tones to 233 million tones – more than four-fold increase.
- The various components of cereals production indicate that whereas cereals accounted for 84 percent in food grains in 1950-51, their share has increased to 94 percent in 2001-02, but as against them, the share of pulses has declined from 16 percent to just 6 percent during the same period.
- Within cereals, the share of the two superior cereals, rice and wheat which was only 53 percent in 1950-51 has improved to 78 percent in 2001-2002, but as against these two superior cereals, the share of coarse cereals has declined from 30 percent to 17 percent during the same period. The reason may be attributed to regular substitution by the weaker sections in favour of rice and wheat as against coarse cereals consumed earlier.

Net Availability of Cereals and Pulses

The data reveal the annual net imports were of the order of 2.3 million tones during 1951-56; they increased to 5.1 million tones during 1961-66 and reached the peak of 6.4 million tones per year during 1966-71. They declined thereafter to 3.6 million tones during 1971-76. During the next 21 years, food imports were negligible ranging between 0.5 to 1.8 million tones: 1 to 1.5 percent of net domestic production. After 1995-96, India became a net exporter of cereals, in fact, India's exports of cereals during 2000-01 was a record 4.5 million tones.

Data about per capita availability of cereals and pulses indicates an overall improvement in per capita availability of food grains from about 419 grams per day to 451 grams between 1951 and 2001. This indicates nearly 15 percent increase in per capita availability during the 50-year period (1951 to 2001). This has two components – cereals and pulses. The per capita availability of cereals increased from 354 grams to 385 grams. This is an increase in per capita availability of cereals by 9 percent during the 50-year period. But in case of pulses, the per capita availability declined from 65 grams per day during to 29 grams, between 1951and 2001. This indicates a decline of about 55 percent in the per capita availability of pulses. Similarly, the per capita consumption of coarse cereals declined from 116 grams per day in 1950-51 to 90 grams per day in 1997-98: a fall of nearly 25 percent. The Ninth Five Year Plan emphasizing this point rightly mentions: "Reduction in per capita pulse availability may adversely affect protein intake. Coarse grains that are inexpensive can provide substantially higher calories for the same cost. If made available through PDS at subsidized rates, this will be self-targeting and improve calorie intake and reduce "hunger" among the poorest segments of population". Obviously, while moving towards foodgrains security, India has succeeded in terms of cereals, but has miserably failed to increase the production of pulses consistent with the needs of a growing population.

Export of Agricultural Products

Exports of agricultural products are dependant on several factors such as domestic production and consumption, exportable surpluses, consumer preferences, varieties traded, quality, domestic and international prices and availability of infrastructure facilities for storage, post-harvest handling etc. Having regard to the social and economic importance of the agricultural sector, the export strategy of the Government is based on the premise that foreign earnings from this sector should be enhanced thereby leading to higher income for farmers, while taking care to make agricultural products available at reasonable prices to the domestic consumers. The export of agricultural products which were valued at Rs. 24, 626 crore at the beginning of the Ninth Plan (1997-98) have gone up to Rs. 27, 922 crore in 2001-02. The agro-products exported can be divided into three broad categories i.e., export of raw products, semi-processes products and export specific and value-added products. The raw products exported are low in value but high in volume; semi-processed products are of intermediate value and limited volume and export specific and value-added products are of high value but low in volume. The role VKC's in providing and dissemination of information, knowledge, etc. on export on agricultural produce/commodity assumes greater significance and scope in the changing economic environment.

Role of VKCs in Agricultural Exports (Plantation Crops)

The role of VKCs in agricultural exports through co-operative organizations such as Tribal Co-operative Marketing Development Federation of India (TRIFED) and National Co-operative Development Corporation is very crucial. They have made elaborate arrangements for export activities linked with agricultural goods like cereals, oilseeds, spices, coffee, tea, cardamom etc. Hence, they can work out ways and means on how to link-up with the VKCs at the micro level through their local, district, regional, state-level federations.

Agricultural Production and the Role of VKCs

Generally, agricultural production denotes the quantum of agricultural crops produced in the economy (country) during a particular period i.e., one year. India is characterized by a lot of imbalances in the contribution of agricultural commodities and products. An analysis of State-wise break-up of agricultural production in very recent years reveals that the development of agriculture is not evenly balanced and it does indicate differences in production. Due to modernization of Indian agriculture, the production is higher in different regions of the country but largely concentrated in some States like Punjab, Haryana, Western Uttar Pradesh, etc. And as a result, their contribution towards agricultural production has been much higher and significant.

According to economists, agricultural productivity is often assessed by measuring the production of agricultural goods (e.g. the yield of a food crop) and by estimating its value on the market, thus knowing the potential for profits. When to recover the golden day's productivity? When it can be done? What strategy is to be used? Where it should be done primarily? What is the role of farmers in this context? What will be the impact on the agriculture? These are few questions relevant for understanding the agricultural productivity. However, the productivity is 'the capacity to produce' with unit resources. In other words, it tells how much of agricultural production takes place in the economy with one unit of land, capital or labourer; or what is the cost of producing one unit of agricultural unit. Hence, one enormous advantage of VKCs is to provide knowledge on both marketing and intelligence on time, which shall decrease the cost of cultivation of crops and increase the sale value. VKCs can aid in better decisions on time in, crop management, crop production, capital (finance), and labour availability. Agricultural productivity can be viewed from two angles: land productivity or productivity per acre of land, and labourer productivity or what a labourer can produce on land by his labour (dynamic physique and some little bit of mental ability).

Productivity in India is low and poor, but had been declining from olden days. The survey conducted by the F.A.O points out that the rice yields per acre were 50 percent higher during the Mughal periods than at present. Over the 15 years from 1961-62 onwards, the productivity increase was around 29 percent points and this too had taken place after 1966-67 due to the introduction of High-Yielding Varieties (HYVs). From 1970 onwards, there was no significant improvement in productivity. Besides, the progress of productivity during the Plan period cannot be considered very satisfactory when compared to heavy investments incurred for enhanced productivity. VKC is the only way and means to disseminate knowledge of enhancing agricultural productivity.

Consequently, the labour productivity per engaged worker in agriculture is also very low in India. Agricultural labour productivity is ridiculously low in India when compared region-wise and also when compared to other foreign countries. VKCs can be used to provide information and data on the movement of labour from one region to the other part of the country to ensure maximum productivity and reduction of cost of labour.

Earlier Initiatives to Promote VKCs

Village Knowledge Centres (VKCs) was earlier initiated in Pondichery by M.S. Swaminathan Research Foundation (MSSRF) in 1998 and thereafter many other states in India adopted it. It focused on information fostering agricultural and overall rural development through the use of Information and Communications Technology (ICT). It created the idea of evolving the value addition centre (VAC), a sub-station at Villanur Pondicherry connecting ten village knowledge centres through a hybrid wireless network comprising of computers, telephones, VHF duplex radio devices and facilitating both voice and data transfer. Local volunteers, mostly women, gathered the information and fed it into the internet by using indigenous knowledge with generic information including those propounded by the Subject Matter Specialist (SMS) and use of Multimedia Interactive System (MIS).

The VAC is done for VKC by preparing it in the local language (Tamil). It has generated a number of databases to answer the queries of the local people's day-to-day requirements. The advertisements and information are published in the media as rural yellow pages. The facilitator was given adequate education and training and providing a token incentive of Re. 1 per visitor to the centre. This initiative paved way for the growth and development of Village Knowledge Centres.

Pre-requisites for Accelerating the Growth of VKCs

In the context of globalization and new economic scenario, the villagers (farmers) are required to understand and enhance their agricultural productivity, production cost, price and marketing of the produce/commodity. They must improve the quality of their agricultural produce/commodities. As a pre-requisite, they must understand details such as genetic (i.e. farm products), legal (their rights), quality (i.e., sanitary and phytosanitary measures) and commerce, business, trade (i.e., local prices and external markets) knowledge. Village Knowledge Centres (VKCs) play a significant role to measure and dispense local information on good governance, e-governance entitlement and state of the art infrastructure or world class Infrastructure and access to dynamic relevant information on weather, markets and online public services etc. The mission of MSSRF and the present government is to promote knowledge centres in more than 600,000 villages. Moreover, it has been formed to mobilize partnership among the public and private sectors, academicians and professionals, mass media and civil society organizations.

Objective of VKCs

There are several objectives of a Village Knowledge Centre. The major objectives are mainly to provide information and data and promote communication process particularly those which are required and relevant to the agricultural information system in the rural areas. Some of the objectives are:

- To disseminate information directly to the rural communities at the grass root level/micro level.
- To predict rainfall and encourage water harvesting technologies during rainy seasons.
- To bring access to a range of information of services relating to agricultural growth, production cost, demand, supply, marketing and how to increase productivity.
- To disseminate market information system to the rural people, the marketing requirements of their produce or demand and supply caused by changing economic environment.

- To know the consumers' requirements and meet their needs and wants on time.
- To provide information and knowledge to the farmers on the availability of fertilizers, pesticides and other agricultural requisite including adoption new agricultural technologies brought about by technological changes in the new economic environment.
- To promote the rural information infrastructures by developing highly developed rural communication technology.

Consequently, in the changing economic environment, the Village Knowledge Centres are essential to facilitate the flow of rural goods, services and products in the global markets in a more efficient and effective manner. Since the market is growing global, the rural farmers may be required to increase their agricultural productivity by adopting new advanced technology and methods of production and is possible through the setting up of Village Knowledge Centres. The Village Knowledge Centre will provide information, knowledge and communication to the farm men and women who require dynamic information relating to farm management, meteorological and market factors related to areas such as agro-forestry, agro-processing, crops and animal husbandry etc.

VKCs can help to provide economic, social and cultural opportunities to all class of people living in the rural areas. It should provide essential data and information requirement for making decisions by farmers, fishermen, craftsmen, businessmen, unemployed youth, entrepreneurs and students by understanding their own requirement and identify their core competencies. VKCs can empower the youth to upgrade the level of development activities for their villages such as establishing rural enterprise which will facilitate large scale employment. There is a possibility of developing PURA enterprises to take advantage of the PURA Programmes/ Schemes. The government should improve e-infrastructure in village areas through its National e-Governance plan. Through VKCs, it can emphasize on rural IT initiatives to non-formal, non-degree learning etc. The VKCs can be a powerful instrument or tool in concern of right to information (at the correct time and place), content creation and capacity building in the rural areas (Micro-level ICT strategies).

Scope and Significance of VKCs

The scope and significance of VKCs can be understood in following terms:

1. VKCs can efficiently enhance the procurement, transportation, processing, marketing and distribution of agricultural products/commodities from rural to urban and from national to globalized market environment.
2. The emergence of VKCs and info-kiosk movement can be promoted by NGOs and corporate sector. The local panchayats and SHGs can access the scientific and technical knowledge, to solve the local problem with greater care. This has been done by Indian Tobacco Company (ITC), Godrej, Technomind, etc. in some small scale manner, hence, needs enlargement of scope and dimensions in the changing economic environment.
3. Through National Digital Gateway for rural livelihood security, it can monitor and communicate climate changes to farmers who can take effective decisions regarding the kharif and rabi seasons. This is very important due to the global warming problems caused by the Green House Gas (GHG) emission which causes negative effects on environment and ecology.
4. The rural ICT network helps in farming viability, sustainability and scalability of the rural knowledge revolution to improve the lives of rural families.
5. It helps to understand the feasibility of market potential in the agricultural trade movement and agri-business management.
6. It creates sustainable rural micro-enterprises in the area of

agro-forestry, agro-industry, fisheries, handicrafts, animal husbandry, poultry farms, rural dairy farms, etc.

7. It helps not in farmer's knowledge regarding the agriculture, but also the knowledge about the current agricultural affairs and present conditions of the economy in the changing economic environment and global scenario.
8. It also provides a vision to farmers to develop their future generations who as techno Savvy and ICT enabled to improve and enhance agricultural productivity.
9. There is a chance to develop a new economic strategy by the illiterate farmers to become literate through VKCs. Due to this, the Village Knowledge Centre is vital to the farmer's living conditions in the rural areas.
10. It will lead to a new 100 percent literate country, if used properly and adequately by the rural farmers.
11. It is also an influencing factor to make India become a developed country by 2020.
12. It helps to gain higher living standards of the rural farmers and the entire people (community) in the rural areas.

Constraints and Prospects of VKCs

Crucial obstacles to theses initiatives are:

- Lack of local/non-availability of global content in their regional language.
- Physical (e.g. electricity and other electronic equipments) and financial problems in the permanent establishment and maintenance of village level kiosks.
- Lack of organizational and institutional structures.
- Insufficient network between institutions in information management and administration.
- Poor policies/strategies to formulate a world class roadmap for harnessing such ICTs initiatives by addressing the VKC's constraints faced by the institutions and government agencies.
- Inadequate proactive role both in network, content

development and transfer, adoption, diffusion and absorption of technology (Roger's Diffusion Innovation process).

- Deficiency of public investment in VKCs to promote rice, maize and all other crops production. This has impact and effect on the productivity.
- Nature of technology which is used currently is extremely poor and inadequate.
- Predominance of marginal and small holdings in the rural areas and being enormously scattered in several agro-ecological and agro-economic zones.

Organizations for Implementation of VKCs

Many organizations can take up VKCs, such as Self–Help Groups (SHGs), Non-Government Organizations (NGOs), District Administration, Village Panchayats, CAPART, Local Communities, MSSRF, Agricultural Cooperative Marketing Organisations, etc.

Even in the 21st century and changing economic environment, agriculture and allied activities remain the major backbone of Indian economy. Most of the agricultural families (70 percent) and industrial production sector mostly depends on agriculture and agro-industrial business activities. Agriculture is not only considered as a priority sector in the economy, but is also considered a foreign exchange earner due to large exports of items like, oilseeds, pulses, cereals, rice, plantation crops i.e. tea, coffee, spices, cardamom etc. Before Independence, the major natural resources and raw materials had been swallowed by British colonial government through promotion of their European based industry i.e. Manchester in U.K. After Independence, the country achieved a lot of growth and progress in various sectors: even if, other sectors (industry, services sector) organizations are using raw materials and other fundamental resources but when we compare those sectors, agriculture is indispensable. Because of the population explosion taking place in our country, the food and nutritional security is very crucial. Sometimes, natural disasters have directly affected the Indian agriculture and its allied sector, but still it is performing well as compared to other sectors because of the efforts taken by the government and research institutions /organizations such as ICAR etc.

Now, we have to struggle to enhance prosperity, productivity and gain higher achievement and growth in agricultural field, more land being brought under cultivation, irrigation, research and development (R&D) processing, export etc. by linking with agricultural co-operative societies through VKC's. Here, the co-operative sector has exclusively offered its unchallengeable functionaries and direct investment to promote growth and prosperity of Indian agriculture, i.e. through Primary Agricultural Cooperative Banks (PACBs) modern processing co-operatives, and dairy co-operatives etc. Also, we have to introduce new varieties of crop, fruits and cereals through large-scale production of hybrid varieties and scientific research carried out by the well-established agricultural research stations in India through linking with VKC's.

The fast moving world economy towards establishment of a competitive market economy within countries and in between countries have changed the operations and status of agriculture and allied sectors in India. In this competitive millennium, the co-operative sector should attempt to develop new co-operative approach by linking with VKC's in order to enhance the growth production, productivity and development of agriculture sector. There is an urgent need for evolving new co-operative approach and new generation co-operative perspective though VKC's i.e. by starting and initiating new generation co-operative organizations using ICT in order to enhance prosperity, growth, production, productivity and overall agricultural production and improve the socio-economic livelihood of the farmers and at the same time make agricultural sector more sustainable, more productive and more profitable in the 21st Century. All these is possible through effective and efficient linkages operations of VKC's at the grass root with apex national level federations like, NCUI, NAFED, TRIFED, NCDC, NDDB, BDDF, IFFCO, KRIBCHO and also the agencies of the Government of India at the grass root level.

Conclusion

In the changing economic environment, there is an imperative need for immediately empowering rural farmers and community through higher agricultural productivity for prosperity and posterity by adopting ICT strategies at the micro level. This can be done by localization of Information and Communication Technology to attain higher agricultural production and productivity in the 21st century.

References

Agricoop.nic.in/stateatglance2003 htm.

Blakie. M.J. and Dent. J.B. (1978) *"Information Systems for Agriculture"*, Applied Science Publishers, Ltd, Ripple road, Barking, Essex England, UK.

Department of Agriculture, *"Memoirs of The Department of Agriculture, Madras"*, Dept. of Agriculture, Madras, 1949.

Dharam Narain, et.al, (1988) *"Studies on Indian Agriculture"*, OXFORD University press, Delhi.

Meghnad Desai, et.al, *"Agrarian Power and Agricultural productivity in South Asia"*, OXFORD University press. 1984.

N.T, Patel, *"Inputs Productivity in Agriculture"*, OXFORD & IBH Publishing Co. 1982.

Samwel Kakuko Lopoyetum (3, Nov.2004) *"Development and Transformation of Dairy Cooperative Marketing Organisations in the Era of Economic Liberalization and Globalization (WTO)"* Published in "Problems and Prospects of Cooperative Movement: Perspectives For 21st Century" by Patil. J.F. et.al, Department of Commerce and Management, Shivaji University, Kolhapur.

Samwel Kakuko Lopoyetum and Selvamani. P, Nov.2006, *"Globalisation and Cooperative Movement in India"*, Third Concept (International Journal of Ideas), Vol. 20, No.237, pp. 15-21.

Samwel Kakuko Lopoyetum and Velu Raj. R (2005) *"Computerisation and Networking (MIS) of Dairy Cooperative Organisations in Tamil Nadu"* Tamil Nadu Journal of Cooperation.

Samwel Kakuko Lopoyetum, (1999) *"Computer Application in the Management Administration of Cooperatives"*, Tamil Nadu Journal for Cooperation, Vol. 91, No. 4.

Samwel Kakuko Lopoyetum, (1999) *"Indo-Kenya Dairy Cooperative System; Scope for Dairy Cooperative Development and Replicability Strategy of Indian model in Kenya"*, Cooperative perspective, Journal of Cooperative Management. Vol. 34, No. 3.

Samwel Kakuko Lopoyetum, (2003) *"Appropriateness of Management Strategies; Are they myth to Cooperatives"*, The Co-operator, Vol. 40, No. 8.

Samwel Kakuko Lopoyetum, (3, Nov.2004) *"Challenges and Opportunities before Cooperative Organisations and Appropriate Strategies in the New Millennium"* Published in "Problems and Prospects of Cooperative Movement: Perspectives For 21st Century" by Patil. J.F. et.al, Department of Commerce and Management, Shivaji University, Kolhapur.

Samwel Kakuko Lopoyetum, (Dec.2003), *"The Dimension of New Generation of Cooperative-21st century Perspective"*, The Tamil Nadu Journal of Cooperative, Vol. XI, No. 2, pp. 11-14.

Samwel Kakuko Lopoyetum, (July 2005) *"Global Business Management; Strategies for Cooperative Business Organisation in 21st Century"*, Tamil Nadu Journal of Cooperation, Vol. 5, No. 9, pp. 26-29.

Samwel Kakuko Lopoyetum, et.al, (April-2003), *"Fundamental Management Dimensions of Cooperatives; An Analysis of major imperative issues"*, Indian Cooperative Review, Vol. 40, No. 4, p. 245.

Samwel Kakuko Lopoyetum, et.al, *"Essential management strategies for cooperative enterprises and Organisations in the era of open market economy"*, The Co-operator, Vol. XI, No.3, pp. 88-91.

Sankaran. S, "*Indian Economy*", Marghan Publications. 2005.

www.icar.org.in.

www.idrc.ca/pan/telecentre.html.

www.indiastat.com.

www.linx.Best.com.
www.nbu.ac.in.
www.ncss.com.
www.nemoc.navy.mil.
www.search4i.com.
www.wikipedia.com

8

Web-Education and Farmers: An Experimental Study

P. Mooventhan and H. Philip

"Access to Internet will soon be universal, and it can provide unrestricted low-cost access to information, as well as highly interactive distance learning. The Internet will not only facilitate interactions among researchers but also quickly improve their ability to communicate effectively with the potential users of their research knowledge".

Prof. M.S. Swaminathan

Introduction

The world is indeed a small place. This saying stands correct in today's ever growing and changing world order. It would not be wrong to say that today's world is the world of information and telecommunication. Everyday, new technology and inventions are being made in the area of information processing and travelling. There is hardly any area which has not been affected by this. Due to all this, the word 'distance' sounds ironical in present day context. The whole world is changing into a small place where any information can be exchanged by people within few seconds and that, too, in proper and effective way without any loss of data while it is being processed. At the one hand, all these different ways of telecommunication and information exchange have highlighted the necessity for multi purpose development and growth of information technology and on the other, the easy access and use of it has boosted the network of information exchange. All this has been possible through information technology like telephone, computers, Internet, E-mail, multimedia, and web-education technology.

Web-Education Technology

"Educating the people through web"– this process is called web-education. The World Wide Web provides new opportunities and hopeful future for distance education over the Internet and distance education is becoming a widespread used for quite a long time now through newspapers, radio and television. Nowadays the World Wide Web (WWW) is playing a vital role in distance education and e-learning since the networking technologies are emerging very rapidly in the media dominant era. The connectivity of the internet to different kinds of people around the world is increasing steadily, like television and radio. The World Wide Web and other Internet-based collaborative modern communication technology tools have significantly enhanced the ability to train and educate people electronically. The Web-education provides a significant new functionality in transmitting the valid information to the learners and providing open forums for exchange of views and ideas within a short period of time.

Web-education technology is the greatest communication medium because of its popularity and its reach. When integrated with modern communication technology tools such as, online discussion groups, Usenet newsgroups, and Video teleconferencing the web-education can greatly increase learner's level of involvement, perception and retention on learning process.

The web-education provides an effective mechanism for integrating many of these tools into a single interface and is an ideal tool for (rapidly shifting) information which is itself. The Web-education is revolutionizing some areas of study through increased opportunities for learning and alternative formats for information.

Information provided through new communication technologies such as modern information technology gadgets should be need based, demand driven, site specific and should be in local language. With this background the present experimental research study entitled, "Impact of Web-Education on Knowledge and Symbolic Adoption of Farmers – An Experimental Study" was planned. The following specific objectives were formulated.

1. To study the profile characteristics of agricultural web-education learners in selected district of Tamil Nadu.
2. To study the knowledge levels of agricultural web-education learners.
3. To study the symbolic adoption of agricultural web-education learners.
4. To find out the relationship between personal and socio-psychological factors of the learners with the knowledge gain and symbolic adoption.
5. To find out the problems faced by the agricultural web-education learners.
6. To find out the suggestions of the learners about agricultural web-education.

Research Methodology

This study was conducted in Coimbatore and Trichy districts of Tamil Nadu, because these two districts had close proximity with agricultural institutions such as Tamil Nadu Agricultural University, Coimbatore; Forest College and Research Institute, Mettuppalayam and Anbil Dharmalingam Agricultural College and Research Institute, Trichy respectively. From these districts, the list of villages provided with computers and internet facilities were obtained from collectorates. Among the villages six villages *viz.,* Odanthurai, Chikkarampalayam and Marudur from Coimbatore district and Ediyapatti, Anthanallur and Navallur Kuttapattu from Trichy district were selected at random. From each of these villages, thirty respondents are selected with the use of simple random sampling. The respondents are expected to have a minimum educational qualification of 5th standard and above. They will have to undergo pre and post exposure knowledge test.

Abstract of Sampling Procedure

- Number of districts : 2
- Number of villages : 6 (@ 3 villages/district)
- Number of respondents : 180 (@ 30 respondents/village)

Findings and Discussion

This section forms the crux of this report. The data collected as per the research methodology explained in previous chapter have been analyzed and results obtained are discussed. The data were classified, tabulated and then treated with scientific methods to get the results.

Table 8.1: Distribution of Respondents According to Their Utilization Behaviour of Websites

S.N	Category	No.	Percent
1.	**Web sites**	**(N=180)***	
a.	Not utilized	140	77.78
	Utilized	40	22.22
II	**Purpose of use**	**(n=40)***	
a.	Agriculture	11	27.50
b.	Non-Agriculture	29	72.50
III.	**Mode of use**	**(n=40)***	
a.	Browsing	40	100.00
b.	E-mail	40	100.00
c.	Chatting	8	20.00
IV.	**Frequency of use**	**(n=40)***	
a.	Daily	0	0.00
b.	Once in two days	9	22.50
c.	Once in a week	31	77.50
V.	**Message storability**	**(n=40)***	
a.	Hard disc	25	62.50
b.	Hard copies	11	27.55
c.	Floppy / CD	4	10.00

(* Multiple response obtained)

From Table 8.1, it is inferred that about four-fifth (77.78

percent) of the farmers had not utilized agricultural web sites for the betterment of farming; remaining 22.22 percent of farmers utilized the agricultural web sites.

Purpose of Utilization

Table 8.1 shows that two-third of the farmers (72.50 percent) had internet connection for non-agricultural usage, followed by 27.50 percent of the farmers for agricultural purposes. It could be inferred from the table that, it is one of the modern communication tool for sharing information within or outside the country. In future, there will be a possibility to have more agricultural websites and in turn this may motivate the farmers to utilize the content.

Mode of Use

In Table 8.1 it is seen that all those farmers who had access to web sites preferred the mode of browsing and E-mailing. About one-fifth of the respondents (20.00 percent) used chatting mode. It could be concluded that highly literate (collegiate education) farmers only used the chatting mode to convey information to others through internet.

Frequency of Use

Table 8.1 indicates that among the web site users, about two-fifth (77.50 percent) of the farmers were found to use the website once in a week followed by 22.50 percent with once in two days and none of them used it daily. It could be inferred that the farmer browsed website mainly to get information on market related and weather forecast on internet might have been the reason for browsing once in a week and once in two days.

Message Storability

As seen in Table 8.1 majority (62.50 percent) of the farmers were saving or storing the web site information in hard disc (computer itself) followed by 27.55 percent as hard copies (print out) and the remaining 10.00 percent in floppy or compact disc. The information such as market price, correspondence and important documents were taken as printouts.

Relationship and Influence of the Independent Variables towards Knowledge Gain and Symbolic Adoption

Simple Correlation Coefficient: In order to understand the association between the selected characteristics of the farmers and their knowledge gain and symbolic adoption on TNAU website, simple correlation was computed and the results are furnished in Table 8.2.

Table 8.2: Relationship of Independent Variables towards Knowledge Gain and Symbolic Adoption

(N=180)

Variable No.	Variables	Knowledge Gain	Symbolic adoption
		'r' value	'r' value
X_1	Age	-0.434**	-0.120 NS
X_2	Educational status	0.577**	0.217 NS
X_3	Occupational status	-0.068 NS	0.262 NS
X_4	Farm status	-0.428**	0.522**
X_5	Farming experience	-0.147 NS	-0.472**
X_6	Annual income	-0.217 NS	-0.187 NS
X_7	Mass media exposure	0.572 **	0.221 NS
X_8	Computer training undergone	0.184 NS	-0.276 NS
X_9	Computer utilization behaviour	-0.076 NS	-0.216 NS
X_{10}	Innovativeness	0.086 NS	0.452**
X_{11}	Awareness about agricultural websites	0.185 NS	-0.124 NS
X_{12}	Utilization of agricultural websites	-0.016 NS	0.127 NS

NS - Non Significant

** Significant at 0.01 level of probability

* Significant at 0.05 level of probability

Relationship of Independent Variables towards Knowledge Gain

It is observed from the Table 8.2 that out of twelve characteristics studied, two characters were found to be having a positive association with knowledge gain. The characters namely educational status and mass media exposure were found to be positively associated with knowledge gain at 1 percent level of probability. Age and farm status, were found to have a negative association with knowledge gain at 1 percent level of probability.

The other variables namely farming experience, occupational status, annual income, computer training undergone, computer utilization pattern, innovativeness, and awareness of agricultural website and utilization of agricultural website were found to have no association with knowledge gain.

It is an accepted fact that with higher education and mass media exposure of the respondents might have helped to get more knowledge gain. The variable age and farm size were having negative association with knowledge gain. The increase in age might have led to less learning capacity of the respondents. A similar trend of increased farm size would have decreased the knowledge gain.

Relationship of Independent Variables towards Symbolic Adoption

The results of correlation analysis of twelve independent variables on symbolic adoption are presented in Table 8.2. It could be explicit from the Table 8.2 that out of twelve characteristics studied; only two characteristics were found to be having a positive association with symbolic adoption of technologies recommended in TNAU website. The characteristics namely farm status and innovativeness were found to be positively associated with symbolic adoption at 1 percent level of probability. Whereas, farming experience was found to have a negative association with symbolic adoption.

The remaining variables age, educational status, occupation, annual income, mass media exposure, computer training undergone, computer utilization behaviour, awareness about agricultural web sites and utilization of agricultural web sites were found to have a non-significant association with symbolic adoption. Larger farm size and more acreage under cultivation might have increased the symbolic adoption level in order to try new technologies in their farm.

More innovativeness of the farmers, greater would be the symbolic adoption of technologies recommended in TNAU website, it might be due to the probable reason of interest in getting additional yield and more returns from the farming. Further the innovators are always ready to accept any new technologies. This could be possible reason for having positive relation.

Higher the level of experience in farming would have decreased the symbolic adoption due to rich experience in farm maintenance and so on. This finding makes it clear that, those who had rich experience in farming were exposed to mass media and other sources of information such as farm scientists and extension personnel to gain knowledge about the recent technologies.

Influence of the Independent Variables towards Knowledge Gain and Symbolic Adoption

Multiple Regression analysis: The actual relationship between measured variables in social science are by no means as simple as above one variable might be associated with or dependent upon several other variables at the same time and hence the multiple regression analysis was carried out, the result of which are furnished in Table 8.3.

Influence of the Independent Variables towards Knowledge Gain

From the Table 8.3, it could be seen that the coefficient of multiple determination *viz.,* R^2 was 0.6127, which was significant at 1 percent level. This meant that 61.27 percent of the variation in the dependent variable is explained by the independent variables chosen for the study. The partial regression coefficient value was found to be negatively significant for the variable, age at one percent level of significance and the farm status behaviour at 0.05 percent level of significance. This revealed that a unit increase in age would decrease the knowledge gain by 0.198 units *ceteris paribus*.

Table 8.3: Influence of the Independent Variables towards Knowledge Gain and Symbolic Adoption

(N=180)

Variable No.	Variables	Knowledge gain	Symbolic adoption
		't' value	't' value
X_1	Age	-2.681**	0.162 NS
X_2	Educational status	2.786**	0.176 NS
X_3	Occupational status	0.087 NS	0.390 NS
X_4	Farm status	-1.982*	2.665**
X_5	Farming experience	0.812 NS	0.610 NS
X_6	Annual income	-0.086 NS	-1.471 NS
X_7	Mass media exposure	3.498**	-2.812**
X_8	Computer training undergone	0.525 NS	1.420 NS
X_9	Computer utilization behaviour	-0.086 NS	-1.781 NS
X_{10}	Innovativeness	0.628 NS	3.217**
X_{11}	Awareness about agricultural websites	-1.567 NS	-0.056 NS
X_{12}	Utilization of agricultural websites	-0.087 NS	-0.467 NS

R^2 = 0.6127 (knowledge gain)
R^2 = 0.4012 (symbolic adoption)
NS - Non Significant
** Significant at 0.01 level of probability
* Significant at 0.05 level of probability

Farm status had negative coefficient which means that a

unit increase in farm status, *ceteris paribus*, would result in decrease of knowledge gain by 0.420 units. The variables educational status and mass media exposure had influenced positively and significantly to knowledge gain. This showed that a unit increase in the educational status and media participation, *ceteris paribus* would increase the gain in knowledge by 0.24 and 0.32 units respectively.

Influence of the Independent Variables towards Symbolic Adoption

The multiple regression analysis was carried out to assess the extent of influence of twelve independent variables towards symbolic adoption. From Table 8.3 it could be observed that coefficient of multiple determination *viz.,* R^2 was 0.4012 which was significant at 1 percent level. This meant that 40.12 percent of the variation in dependent variable is explained by the independent variables selected for the study.

Table 8.3 vividly shows that farm status and innovativeness had contributed positively at one percent level of significance towards symbolic adoption whereas mass media exposure contributed negatively at one percent level of significance. A unit increase in farm status, and innovativeness, *ceteris paribus,* would result in increasing the symbolic adoption to an extent at 2.16 units and 0.07 units respectively.

The knowledge aspects of the selected subject matter areas (TNAU website) were exposed to the six experimental groups (Group I, Group II, Group III, Group IV, Group V and Group VI) through single treatment. The knowledge levels of the subjects before and immediately after exposure were assessed to find out the knowledge gain. The maximum attainable knowledge score was 31. The result of the selected groups knowledge gain is presented in Table 8,4.

It could be seen from Table 8.4 that all the groups namely Group I, Group II, Group III, Group IV, Group V and Group VI had effective in terms of knowledge gain. From the result, in Coimbatore district the Group I (Maruthur) had highest (60.00 percent) knowledge gain followed by Group II (Chikkarampalayam) 59.16 percent and Group III (Odanthurai) 42.66 percent and from Trichy district the Group IV (Anthanallur) had maximum (53.53 percent) knowledge gain followed by Group V (Ediyapatti) 53.53 percent and Group VI (Navallur Kuttapattu) 40.06 percent. Among the six groups, four groups namely Group I, Group II, Group IV and Group V had no significant different in knowledge gain. But the Group III and Group VI had minimum knowledge gain (42.66 percent) and (40.06 percent) respectively. Even through these groups got maximum score in the pre exposure knowledge test. It might be due to the fact of their close proximity with Forest College and Research Institute and Anbil Dharmalingam Agricultural College and Research Institute respectively.

Table 8.4: Mean Knowledge Gain due to Exposure to the Treatment (N=30 each group)

S. N	Groups	Name of the Villages	MKG		MKG	%
			Pe	Poe		
	Coimbatore District					
1.	Group I	Maruthur	4.12	22.12	18.00	60.00
2.	Group II	Chikkarampalayam	3.52	21.27	17.75	59.16
3.	Group III	Odanthurai	8.92	21.72	12.80	42.66
	Trichy District					
4.	Group IV	Anthanallur	4.12	21.07	16.95	56.53
5.	Group V	Ediyapatti	3.42	19.98	16.00	53.53
6.	Group VI	Navallur Kuttapattu	7.99	21.72	12.02	40.06

MKG = Mean knowledge gain; Pe = Pre-exposure; Poe = Post-exposure;

From the Table 8.5, it is inferred that about 47.77 percent of respondents were ready to adopt alternative cropping such as sweet sorghum, sugar beet, and jatropha as per the recommendation of TNAU website followed by 53.89 percent of respondents who were ready to adopt crop management practices, 54.44 percent of respondents were ready to adopt agricultural produce processing technologies, 61.11 percent of respondents were ready to adopt chemical herbicides and 66.66 percent of respondents were ready adopt post harvest technologies.

Table 8.5: Effectiveness in Terms of Symbolic Adoption

(N=180)

S.N	Questions	Number*	%
1.	Do you propose to use the alternative crops like sweet sorghum, atrophic, sugar beet recommended by TNAU website?	86	47.7
2.	Do you propose to use the recommended crop management practices recommended by TNAU website?	97	53.8
3.	Do you propose to use the agricultural produce processing technologies recommended by TNAU website?	98	54.4
4.	Do you propose to use the chemical herbicide recommended by TNAU website?	110	61.1
5.	Do you propose to use the post harvest technologies recommended by TNAU website?	120	66.6
6.	Do you propose to use the organic farming procedures recommended by TNAU website?	126	70.0
7.	Do you propose to use the chemical pesticides recommended by TNAU website?	130	72.2
8.	Do you propose to use the agricultural farm machineries recommended by TNAU website?	137	76.1
9.	Do you propose to use the course material related to agricultural technologies recommended by TNAU website?	146	81.1
10.	Do you propose to use the fertilizers dose recommended by TNAU website?	147	81.6
11.	Do you propose to use the web-education course material recommended by TNAU website?	154	85.5

(*Multiple response obtained)

Further nearly 70.00 percent of respondents were ready to

adopt organic farming procedure, 72.22 percent of respondents were ready to adopt chemical pesticides, 76.11 percent of respondents were ready to adopt farm machineries, 81.11 percent of respondents were ready to use course material related to agricultural technologies, 81.66 percent of respondents were ready to adopt recommended fertilizers and most of the respondents were ready to adopt the technologies given in web-education course material for specific crops as per the recommendation of TNAU.

Problems in Utilizing Web-Education Technology

It is necessary to study the constraints in any social research. The constraint analysis helps the planners, administrators, development workers, scientists and others to frame policies and to implement schemes. Hence, the constraints faced by the farmers in utilizing web-education technology were studied. The multiple responses of constraints were worked out. The results are presented in Table 8.6.

The result could be observed from Table 8.6. The major problem as expressed by all respondents (100 percent) was lack of information in local language and this was due to the fact that most of the web information published in English, high cost for establishment (95.55 percent), illiterate people cannot use (93.88 percent), difficulty in handling of computer mouse (91.66 percent) and this might be due to lack of practice and proper training on the use of mouse in computer and its accessories, difficulty to type the URL to access the web page (88.86 percent) was expressed because all the URL addresses are exclusively in English language, high cost to connecting internet (87.22 percent) and this was due to the reason of unavailability of internet providers at village level, lack of skill to use modern information technology gadgets (78.88 percent) as most of the respondents lacked training on the use of modern gadgets, lack of familiarity (77.77 percent), lack of relevant information (73.33 percent), lack of prompt reply to online queries and clarification (53.88 percent), inability to make or receive electronic payments (50.00 percent), slow downloading and uploading of photos/files (45.55 percent), lack of update information (41.66 percent) and few percent of the respondents expressed that they had difficulty in reading online information.

Table 8.6: Problems Expressed by Web-education Learners

(N=180)

S.N	Problems	Number*	%
1.	Lake of information in local language	180	100.00
2.	High cost for establishment	172	95.55
3.	Not suitable to illiterate people	169	93.88
4.	Difficult in handling of computer mouse	165	91.66
5.	Difficulty to type the URL to access the web page	160	88.88
6.	High cost for internet connection	157	87.22
7.	Lack of skill to use modern information technology gadgets	142	78.88
8.	Lack of familiarity towards modern information technology gadgets	140	77.77
9.	Lack of relevant information in the website	132	73.33
10.	Lack of prompt reply to online queries and clarification	97	53.88
11.	Inability to make or receive electronic payments	90	50.00
12.	Slow down loading and uploading of photos/files/ videos	82	45.55
13.	Lack of update information	75	41.66
14.	Difficulty in reading online information	62	34.44

(* Multiple response obtained)

Suggestions for Effective Utilization of Web-Education Technology

Any study would be complete if it gives suggestions for development in the field of investigation. The suggestions rendered by the respondents for effective utilization of web-education technology were studied. The results are presented in Table 8.7.

Table 8.7: Suggestions for Effective Utilization of Web-education Technology (N=180)

S.N	Suggestion	Number*	%
1.	Information may be in local language	180	100
2.	Location specific information may be provided	172	95.55
3.	Training may be provided to learn computer in detail	170	94.44
4.	Regular update of website information	167	92.77
5.	Installation of software in Tamil of word, excel, power point, e-mail can be installed in information kiosk	160	88.88
6.	Valarum vellanmai may be published in online	157	87.22
7.	Information kiosks may provide location specific weather data	152	84.44
8.	Linking of Village Information Centre with State Agricultural University, Krishi Vigyan Kendra and development departments	149	82.77
9.	On line marketing facilities for farm produce	137	76.11
10.	Keeping the agricultural website addresses on the desktop of computers in Village Information Centre	102	56.66
11.	Self Help Groups can run the information centre	91	50.55
12.	More number of information kiosk	71	39.44
13.	Use of video conferencing	51	28.33

(* Multiple response obtained)

Suggestion rendered by the respondents could be seen from Table 8.7. Major suggestions were giving information in local language expressed by all the respondents (100.00 percent) for easy comprehension by all farmers, followed by location specific information (95.55 percent) to reach all farmers who are spread across the state and country with diversified climatic condition, training to operate computers (94.44 percent) due to digital divide among rural farmers on the usage of computer, regular update of website information (92.77 percent) to fulfil the latest and current agricultural information need of the farmers, software in Tamil of word, excel, power point, e-mail may be installed in information kiosk (88.88 percent), valarum vellanmai may be published in online (87.22 percent), information kiosk may provide location specific weather data (84.44 percent), linking of Village Information Centre with State Agricultural University, Krishi Vigyan Kendra and development departments (82.77 percent), on line marketing facilities for farm produce (76.11 percent), keeping the agricultural website addresses on the desktop of computers in Village Information Centre (56.66 percent), Self Help Groups may run the information centre (50.55 percent), more number of information kiosk (39.44 percent) and very few percent of the respondents suggested that use of video conferencing is required for effective utilization of web education technology.

Conclusion

The study revealed that the non-computer owning respondents lack skill to operate computer and to know about technologies given through web-education. So information kiosks should be installed in villages and skilled trainers should give proper and regular trainings. The results clearly depict that majority of the respondents were unaware of website-education, despite the fact that they are more interested to know about it. Such large scale awareness campaigns and mass media can be utilized in a big way to promote web-education technology. The study revealed that respondents were satisfied with the tool web-site but not convinced with the technology existed on it. So the scientific community should take initiatives to evolve technologies, which would compromise farmers in terms of affordability and effectiveness. The study revealed that respondents are more satisfied with the web-education and transfer of technology through computers.

Further, this finding has indicated that they are inclined towards accepting Cyber Extension. So the farm graduates who are unemployed can be manned in Information shops and can facilitate farmers to know timely, accurate information through World Wide Web. The awareness level and utilization behaviour of websites was found to be low and this indicated that there is an urgent need for development and popularization of such websites. The variables such as educational status, mass media exposure, innovativeness, farm status and farming experience were found to act as critical variables.

So, while preparing web-education in future, one should take care of above variables. Public and private computer training institutions and entrepreneurs may be entrusted with the task of opening information clubs on partnership with farm association to teach, train farmers and keep away their fear in using this new technology in rural areas. This may be done as a Government policy initiation.

References

Anandaraja, N. (2002). "*Developing Farmer Friendly Interactive Multimedia Compact Disc and Testing its Effectiveness in Transfer of Farm Technology*". Unpub. Ph.D. Thesis, TNAU, Coimbatore.

Karunakaran, B.(2004). "*A Study on the Potential of Modern Information Technology Gadgets for Agricultural Development*". Unpub. M.Sc. (Ag.) Thesis, TNAU, Coimbatore.

Senthilkumar, M. (2003). "*Field Testing Cyber Extension Techniques for Transfer of Farm Technology-Feasibility Study*".

Unpub. Ph.D. Thesis, TNAU, Coimbatore.

The Chronicle Yearbook, (2005). "*Information Technology*". Chronicle publication private Ltd. New Delhi.

9

Knowledge Mission 2007 and Rural Farm Women

A. Thomas William and G. Albin Joseph

"The life blood of digital revolution is freedom" - *Kofi Annan.*

The World Summit on the Information Society (WSIS) held at Turin on November 15 2005 visualized to 'Connect the World by 2015'. Its aim was to provide and ensure the benefits of the digital revolution to reach every country and every part of the each country by the year 2015, which is also the benchmark year for achieving the UN Millennium Development Goals. Information and communications technology (ICT) has been with us for many years, and it has played an imperative task in promoting agriculture and rural development during the last several decades.

Knowledge Mission 2007

The ultimate goal of Knowledge Mission 2007 is to 'Decrease the gap of Rural Urban Digital Divide' and provide knowledge connectivity to every village of India by August 15, 2007. In the light of the above mission, a National Alliance has been established, involving 22 Governmental organisations, 94 civil societies, 18 academic institutions and 10 financial institutions and–still more to join in the course of the time. The main characteristics of a knowledge society is one that empowers people and communities, increasing the effectiveness of their development efforts through informed decision-making and through their capacity to harness science and various forms for knowledge to achieve the objectives for poverty eradication, food security and sustainable development

(Ghosh, 2001).

Recent Developments in the Mission 2007

1. One lakh ICT-based community service centres or Village Knowledge Centres (VKC) by 2007 by the Department of Information Technology, Government of India. Leveraging SWAN (State Wide Area Network) infrastructure, community service centres will provide reliable broad-based connectivity to remote villages.
2. Setting up of Village Resource Centres (VRC) at the block level by the Indian Space Research Organization in collaboration with appropriate public and civil society institutions to provide a wide range of services including tele-conferencing facilities.
3. The decision of the Ministry of Panchayati Raj to establish Internet connected ICT centres in all the 2,40,000 panchayats and local bodies in the country by August 15, 2007.
4. The Rural Information Society Initiative (RISI) of BSNL will aim to set up 1,00.000 VKCs each covering a population of 2000 or more.
5. Promotion of e-governance.
6. Inclusion of e-health facilities under the National Rural Health Mission by the Ministry of Health and Family Welfare.
7. Setting up of Public tele-information centres (PTIC) through its Universal Service Obligation Fund (USOF).
8. Enactment of Right to Information Act, 2005,

These ICT initiatives should be designed in such a way that it should be:

- Pro-poor,
- Pro-women, and
- Pro-livelihood.

The major requirement for the success of VKC's is dependent on the sense of ownership, content creation and capacity building (Swaminathan, 2005). The world needs the information technology to unleash the true potential of its people without denying to the people the full access on political, social and other grounds. Hence, the lifeblood of the digital revolution is freedom. (Annan, Kofi 2005) The information revolution in India could be witnessed during the Turin Summit 2005. (Parthasarathy, 2005)

It is to be ensured that the powerful tool in the form of new information technologies are used to guarantee, in the words of the UN Charter, 'better standards of life in larger freedom'. Greater access to information and communications technology, or ICT, can improve the lives of farmers and assist micro-entrepreneurs. It can prevent AIDS and other communicable diseases, promote women's equality and foster environmental protection. Indeed, all over the developing world, electronic commerce, distance education, telemedicine and e-governance are already improving the quality of life for countless people. (Tharoor, 2005)

Need for the Study

There is still a content divide which would doubt that media principally reflects the interest of their producers. And whether we look at television, radio or the Internet, what poses for global media is really the media of the developed countries in the West. Most of the globe's Internet hosts are based in industrial countries and most of the content is sill produced in and for, Western countries, in Western languages. (Tharoor, 2005)

The relevance and timeliness of the content will determine the interest of rural families in VKC's. The content should be demand driven and area, culture and time specific (Swaminathan, 2005); otherwise locale specific. The various stakeholders have to work synergistically and in mutually collaborative manner.

The National Alliance has suggested that at the level of each district, a 'Content Consortium' may be organised to enable VKC managers to access the information they need in the area of weather, health, entitlements to government projects, e-governance, credit and insurance, agricultural goods marketing. The managers of VKC's will have to maintain active contact with the content consortium. To facilitate the availability of the right information at the right time and place, it is proposed to organise 'National Digital Gateways' for agriculture, education, health, and livelihoods.

Role of Rural Women in Agriculture

Women play a pivotal role in agriculture. The rural farmwomen are often physically visible but conceptually invisible and remain marginalized. The best way to make the optimum use of human resources is to provide them adequate information for self-development and improve their existing knowledge and skill so that their capacity could be enhanced. Women participate in significant numbers in all agricultural activities. The extent to which these activities are recognized and given their due, depends on, among other factors, socio-economic-ethnic status, customs and traditions, religious beliefs and education. (Anuva, 2001)

Women in rural India are involved in various activities starting from home to land. They play an important role either a predominant one or a subsidiary one according to the situation, need, nature of the husband, culture of the family and so on. Their activities range from crop production, animal and poultry farm management, household and familial jobs and participation in social development of the village. However, over a period of time due to change in technological advancement, the role of women has undergone a significant change. It is also important to keep in mind the fact that women also spend a substantial part of their time in the management of livestock, poultry, allied activities, etc. (Sisodia, 1985; Agarwal, 1988; Sethi, 1989)

Women contribute a significant share of the labour use in crop production and also spend considerable amount of time in livestock, food processing, sericulture and weaving activities.

These time commitments are in addition to the amount of time spent in household chores. (Anuva, 2001)

Study Area

The study area is Thirumangalam Taluk in Madurai District, Tamil Nadu. The total area is 814.18 sq. Km. with a total population of 1.95 lakh (2001 census) comprising of 97 thousand each, both male and female. The total literates in this taluk are about 1.27 lakh. Thirumangalam is a Panchayat Union consisting of 38 Village panchayats. Except Thirumangalam, all the other areas are villages and hamlets. The major occupation is agriculture and allied activities. The other unorganized sectors are, construction work, manual work, petty business, head load business, sheep and goat rearing etc.

Methodology

120 farm women of Thirumangalam block of Maduari district of Tamil Nadu comprising of 10 villages were selected by stratified random sampling method. The primary statistics on the list of women, either small farmers or marginal farmers were obtained from leading NGOs in the area. Only women belonging to the small and marginal farmers' families were included for the study.

Data was collected from the selected households where women had participated in the farming activities. Specially designed interview scheduled was prepared in a prescribed proforma by surveying 120 households. Data is collected from 10 villages comprising 6 each from small farmers (1.0–2.0 ha) and marginal farmers (below 1.0 ha).

Specific list of information needs for each area is carefully prepared and pre-tested. Three-point scale is used to calculate the total and mean score and the rank values are assigned.

The three-point scale was:

- 3 – Mostly needed
- 2 – Needed

- 1 – Less needed

Objectives

- To identify the information needs in the area of cultivation practices, animal and poultry and allied activities.
- To assess the important and urgent information needs on family and household activities and on the social and development aspects.

Major Observations

In Table 9.1, 14 important areas related to farm operations by women were identified and mean score was calculated to identify the major ones.

The rural women identified the need to know the marketing details regards to price, place, trend and related issues as the predominant and most important information content to be included in the content creation. Thus, the access and availability on information on marketing of agricultural products is lacking for rural farm women.

The method and procedure, the quantity, concentration and time space in application of manure and fertilizer is observed as the second important farm operation need required for the rural farm women with the mean score of 2.91.

The next equally important area where the information is needed to be included in the content creation is the pest and disease management (Mean score 2.90).

The other areas (Table 9.1) to be covered in the content bank are irrigation techniques (2.83), intercropping (2.64), and meteorological information (2.60) with regards to their local area.

In Table 9.2, 15 important needs on animal, poultry and allied activities were identified and the women were asked to score, ranked as per the importance to them.

The following observations are made:

Care for the younger ones finds a prominent place among the needs to be included in the content creation list with the mean score of 2.91 followed by disease. Preventive measures for birds and animals were required to be known to these farm women as this information was lacking. Again information on marketing (2.84) of farm animal products such as milk, egg, meat, etc. with latest details and current trends and the agencies involved and prices are the other details required to be available to the women farmers in the knowledge creation.

Table 9.1: Information Needs on Farm Operations

S.No.	Information Needs on Farm Operations	Mean Score
1	Land and soil conservation	1.60
2	Crop selection	1.94
3	Manure / fertilizer application	2.91
4	Weed management	2.23
5	Meteorological information	2.60
6	Pest and disease management	2.90
7	Irrigation techniques	2.83
8	Labour saving mechanisms	2.55
9	Storage of produce	1.76
10	Marketing of products	2.94
11	Harvesting techniques	1.96
12	Seed selecting and treatment	2.20
13	Information on intercropping	2.64
14	Information on dry land farming	2.50

Mean of Mean score: 2.39

Medical and treatment information (2.80) is the next important need to be met out through the village knowledge centre. Financial institution for loan and subsidy (2.76), income generation during off season (2.76) and their details are expected to be included in the content to be created for ICT terminals. The mean of mean score for all other information criteria related to the caring of farm animals and birds is calculated at 2.33.

Table 9.2: Information Needs on Animal, Poultry and Allied Activities

S.No.	Information Needs on Animal, Poultry and Allied Activities	Mean Score
1	Selection of animals and birds	2.60
2	Housing of animals and birds	2.10
3	Feeding composition	2.40
4	Fodder crop cultivation	1.24
5	Medical and treatment information	2.80
6	Prevention of diseases	2.86
7	Marketing of products	2.84
8	Financial institutions for loan and subsidy	2.76
9	Care of younger ones	2.91
10	Manure and compost preservation and management	1.86
11	Processing and preservation of additional products	1.40
12	Budgeting and accounting	1.54
13	Preservation of vegetables and fruits	2.00
14	Composite and manure preparation	2.68
15	Information on Income generation during off season	2.76

Mean of Mean score: 2.33

The information needs on familial and household aspects are covered in Table 9.3 with 11 key aspects. The following indications were made by the rural women to be included in the content creation.

It is interesting to note that education and employment of children found a predominant place to be included in the knowledge resource with the top mean score of 2.92. It is followed by the data on child care and rearing (2.74).

In the next level, personal hygiene and health is given priority by the rural women. The rural women aspire to have the details regarding the marital life aspects to be covered in the content creation.

The details about the current formation on decision

support system are also required, as the mean score is calculated as 2.43.

Table 9.3: Information Needs on Familial and Household Aspects

S.No.	Information Needs on Familial and Household Aspects	Mean Score
1	Child care and rearing	2.74
2	Nutritional information	1.80
3	Household work	1.60
4	Education and employment of children	2.92
5	Support information for decision-making	2.43
6	Relationship building	1.25
7	Marital life	2.50
8	Network with neighbours	1.50
9	Health and personal hygiene	2.65
10	Budgeting and accounting	1.62
11	Thrift and saving practices	2.20

Mean of Mean score: 2.11

Finally in Table 9.4, the information needs on social and development aspects are covered in 15 relevant areas. The major remarks are presented below.

Recent announcements of government on various issues, changes in the programme or scheme of the government, policy of the government and priority of the government is an essential area which the rural women wanted in the content creation. The next highest mean score of 2.88 is given to the programmes and schemes of the government. How many self-help groups are available in the locality, their activities, functions, success stories and related information are expected to be included in the content creation of ICT.

Government officials at the grass root level–Who is who, his responsibilities and powers, location of the office, the ideal time to meet him, his availability to be made known to the people, is yet another request by the respondents.

Table 9.4: Information Needs on Social and Development Aspects

S.No.	Information Needs on Social and Development Aspects	Mean Score
1	NGO's working in this area	2.40
2	Government development departments	1.85
3	Government officials at the grass root level	2.86
4	Information on PRI representatives	1.63
5	Powers and functions of gram sabha	1.45
6	Role of women in PRI and gram sabha	2.45
7	Local self help group and its activities	2.60
8	Programmes and schemes of government	2.88
9	Recent / latest announcements of the government	2.94
10	General political trend or scenario	1.65
11	Current issues and problems in their village	1.50
12	New varieties of crops and animals	2.20
13	Recent research works of nearby agricultural universities	1.96
14	Research trends in local colleges and universities	1.99
15	Information on block development office	2.55

Mean of Mean score: 2.19

Conclusion

Rural farmwomen contribute their labour, time and expertise to farming activities. They become useful and equivalent partners in the farming activities. The success of knowledge revolution depends on these women and whose information needs are to be met by making the content creation locale specific. The philosophy of ICT in India is pro-poor, pro women and pro-livelihood could only be achieved, when their needs are to be met. The technology could attract women only when their desires are to be fulfilled. Hence, utmost care should be taken to include the information in the areas of farming, allied activities, animal husbandry, familial and household activities and social and developmental activities. A thorough scanning is required on the information needs of various target groups, especially women before creating the content for knowledge centres. It is also equally necessary to update them by research. This paper calls for future research in identifying the information needs of various target groups such as farmers, rural artisans, students, scholars, youth, unemployed youth, children and various other stakeholders.

References

Agarwal, Hema (1988), "*Contribution of Farm Women in Animal Husbandry enterprise",* in T.M. Dak (Ed.) (1988), Women and work in Indian Society, Discovery publishing House New Delhi.

Ghosh, Souvik, (2001), "*Information and communication technologies in rural development, Kurukshetra*, Nov. pp. 2-5.

Giriappa, S; (1988*), "Role of Women in Rural development,* Daya Publishing House New Delhi.

Haffis, Shaik and Y.V.R. Reddy, (2001), "*Women in Agriculture and Rural Development,* Concept Publishing Co, New Delhi.

Parthasarathy, Anand, "*Freedom lifeblood of digital revolution*", The Hindu, 19th November 2005.

Rajula Devi, A.K. (1992), "*Women in Agriculture: A Socio-Economic Analysis,* Concept Publishing Company, New Delhi.

Rani, Usha, D.L/.Vyas and G.S. Hodha, (1993), "*Gender Differential in Work participation in various operations of crop and livestock enterprises in Semi-arid areas of Rajasthan",* Indian Journal of Agricultural Economics, Vol. 48, No. 3, July-Sep, pp. 482-87.

Saikia, A (1985), "*Effect of cropping pattern on Employment of females – A study in sibasagan and Jorhat district of Assam*", Indian Journal of Agricultural Economics, Vol. 40, No, 3. July-September.

Sethi, R.M. (1989), "S*tatus and Power of Rural Women, Women and Development*", Ashish Publishing House, New Delhi.

Sisodia, JS. (1985), "*Role offFarm women in agriculture : A study of Chambal command area of Madhya Pradesh*", Indian Journal of Agricultural economics, Vol. 40, No. 3, July-September.

Sudha, M. et.al (1991), "*A study on the Role of Women in Dry land Agriculture : Utilisation and crop production efficiency",* The Indian Journal of Labour Economics, Vol. 34, No. 3., July-

September, pp. 263-270.

Swaminathan, M.S., "*Mission 2007: Every Village a Knowledge Centre*", The Hindu, 25th Nov, 2005.

Tharoor, Shashi, "*The Great Digital Divide*", The Hindu, 20th November. 2005.

10

Land Use Pattern in Tamil Nadu

P. Mary Elizabeth

Tamil Nadu has a geographical area of 13 million ha which constitutes 3.96 percent of the land area of the country. Land use characterization is critical in government planning and policy strategies. The land use refers to the habitats of vegetation type present, such as forest, agriculture and grassland. The main rivers flowing through the state are the Cauveri, the Pennaiyar and the Vaigai. The total geographical area of the State is 1.30 cores ha. It has been classified according to different types of land use pattern during 04-05 and presented in Table 10.1.

Table 10.1: Land Use Pattern

Land use	Area in ' 000 ha	%
Total Geographical area	13006	
Reporting Area for land utilization	12,998	100.00
Forests	2,141	16.47
Not available for cultivation	2,418	18.60
Permanent Pasture & Grazing land	125	0.96
Land under misc. tree crops & groves	226	1.74
Cultivable waste land	345	2.65
Fallow land other than current fallows	1,229	9.46
Current fallows	1,028	7.91
Net area Sown	5,486	42.21

Source: Land use Statistics at a Glance 1996-97, Ministry of Agriculture, GOI, 2000.

Forests represent all actually forested area on the lands classed or administered as forest under any legal enactment dealing with forests, whether state owned or private. The forest area of the state during 04-05 was 2122069 ha accounting for 16.3 percent of the total geographical area of the state. Erode district with an extent of 228749 ha under forests is the highest among districts in their contribution to the forest area of the state. This works out to 10.8 percent of the state's total forest area. This is followed by Krishnagiri district with 202409 ha (9.5 percent) and Dharmapuri with 164177 ha (7.7 percent).

The unique feature of The Nilgiris district is that about 56 percent of the total area of the district is under forests followed by Krishnagiri with 39.4 percent, Dharmapuri with 36.5 percent, Kanyakumari with 32.4 percent and Theni with 32.0 percent. Erode, Krishnagiri, Dharmapuri, Vellore, Coimbatore, Thiruvannamalai, The Nilgiris, Dindigul, Salem, Thirunelveli and Theni Districts together accounted for 80.1 percent of the total forest area of the state.

Barren and Unculturable Land

Mountains, deserts, hills etc. where cultivation is not possible are classified as barren and unculturable land. An extent of 509275 ha of land comes under this category which represents 3.9 percent of the total geographical area of the state. Villupuram district alone accounted for 56,651 ha which is 11.1 percent of the state's barren and unculturable land and about 7.8 percent of its geographical area under this category. The area under this category is very meagre in Thiruvarur district with just 113 ha.

Land Put to Non-agricultural Uses

The land put to use for purposes other than agriculture such as buildings, pathways, roads, canals, bus stands, railway tracks etc. are brought under this category. Area under this classification is 2124564 ha accounting for 16.3 percent of the state's geographical area. The extent under this category has increased by 11211 ha i.e 0.5 percent as compared to 03-04. In Chennai district, about 98.3 percent (16798 ha) of its geographical area is put to non-agricultural uses because it is a metropolis followed by Kancheepuram district with an extent of 145619 ha (32.9 percent), Thiruvallur with 102036 ha (29.8 percent), Sivagangai with 117367 (28.0 percent) and Pudukottai with 129265 ha (27.7 percent).

Culturable Waste

All lands available for cultivation whether or not taken up for cultivation or taken up for cultivation once, but not cultivated during the current year and continuously for the last five years or more in succession for one reason or the other are classified as culturable waste. The total area under culturable waste is 374026 ha or 2.9 percent of the total geographical area of the state. Karur, Tirunelveli and Thoothukudi districts together accounted for 49.2 percent of the area under this category. The area under this category is nil in Kanyakumari and Chennai districts and very meagre in Erode district with just 527 ha.

Permanent Pastures and Other Grazing Lands

All grazing lands, whether they are permanent pastures or meadows are considered as permanent pastures and other grazing lands. An extent of 113563 ha or 0.9 percent of the geographical area of the state falls under this category. The extent under this category was the highest in Kancheepuram district with 18328 (16.1 percent) ha followed by Karur district with 10801 (9.5 percent) ha

Tree Crops and Groves Not Included in the Net Area Sown

Miscellaneous tree crops such as casuarina, eucalyptus, teak, bamboo, babul etc. which bear yield only once in its whole life span are classified under this category. The extent under this category was 290072 ha i.e. 2.2 percent of the geographical area of the state. Thoothukudi district with 46441 ha ranked first contributing 16.0 percent to the total area of the state under this category followed by Ramanathapuram district with 41116 ha (14.2 percent) and Pudukottai district with 27925 ha (9.6 percent).

Current Fallow Land

Land which is kept fallow during the entire period under report out of the net area sown during the previous year is known as current fallow. The area under current fallow during 04-05 was 691926 ha as against 953963 ha during the previous year with a decrease of 262037 ha i.e. 27.5 percent. The extent under this category is the highest in Villupuram district (92151 ha) followed by Erode (83316 ha), Vellore (71621 ha) and Thiruvannamalai (54974 ha) districts. These districts together accounted for 43.7 percent of the total area of the state under this category.

Other Fallow Land

Land which were taken up for cultivation but have been temporarily put off for cultivation for a period of not less than one year and not more than five years are treated as other fallow land. An extent of 1704139 ha, which is 13.1 percent of the total geographical area, was recorded under this category during 04-05 as against 1862861 ha during the previous year. There was a decrease of 158722 ha (8.5 percent) under this category as compared to the previous year. The land under this category was the highest in Tirunelveli district with 158373 ha and ranked first contributing 9.3 percent to the total area of the state under this category followed by Coimbatore district with 119376 ha, Sivagangai district with 117358 ha and Erode district with 115677 ha during 04-05.

Net Area Sown

Net area sown represents the area sown under first crop during the fiscal year. Out of 13026645 ha of the total geographical area, 5097011 ha of land constituting 39.1 percent was cultivated once (i.e net area sown) with various crops during the year 04-05. When compared to the previous year, the area under this category has increased by 407855 ha (8.7 percent). Of the total net area sown in the State, the share of Coimbatore district was the highest with 324410 ha (6.4 percent) followed by Villupuram district with 320395 ha (6.3 percent)' Erode district with 300065 ha (5.9 percent.) and Dindigul district with 255945 ha (5.0 percent). When comparing the net area sown with that of the geographical area of the district, Thiruvarur district ranked first contributing 69.8 percent of its geographical area towards this category followed by Cuddalore district with 58.6 percent, Perambalur district with 57.6 percent, Thanjavur district with 55.7 percent and Nagapattinam district with 52.7 percent. The net area sown in each district, its percentage contribution to its geographical area and the percentage share of each district to the total net area sown of the state are shown in Table 10.2.

Table 10.2: District Contribution to Net Area Sown (area in ha)

	District	Geographical area	Net area sown	% of net area sown to	
				Geographical area	State total
1.	Chennai	17098		0.0	0.0
2.	Kancheepuram	443210	121362	27.4	2.4
3.	Thiruvallur	342243	99585	29.1	2.0
4.	Cuddalore	367781	215348	58.6	4.2
5.	Villupuram	722203	320395	44.4	6.3
6.	Vellore	592018	172918	29.2	3.4
7.	Thiruvannamalai	631205	235903	37.4	4.6
8.	Salem	520530	202889	39.0	4.0
9.	Namakkal	336335	167151	49.7	3.3
10.	Dharmapuri	449777	169908	37.8	3.3
11.	Krishnagiri	514326	179679	34.9	3.5
12.	Coimbatore	747079	324410	43.4	6.4
13.	Erode	816191	300065	36.8	5.9

Contd...

14.	Tiruchirapalli	440383	171002	38.8	3.4
15.	Karur	289557	100432	34.7	2.0
16.	Perambalur	369137	212765	57.6	4.2
17.	Pudukottai	466329	158724	34.0	3.1
18.	Thanjavur	339657	189295	55.7	3.7
19.	Thiruvarur	209709	146472	69.8	2.9
20.	Nagapattinam	271583	143118	52.7	2.8
21.	Madurai	374173	138560	37.0	2.7
22.	Theni	324230	112053	34.6	2.2
23.	Dindigul	626664	255945	40.8	5.0
24.	Ramanathapuram	408957	191823	46.9	3.8
25.	Virudhunagar	424323	141278	33.3	2.8
26.	Sivagangai	418900	122173	29.2	2.4
27.	Tirunelveli	682308	166419	24.4	3.3
28.	Thoothukudi	459054	179289	39.1	3.5
29.	The Nilgiris	254485	79514	31.2	1.6
30.	Kanyakumari	167200	78536	47.0	1.5
State		13026645	5097011	39.1	100.0

Area, Production and Productivity

Area and production of any crop largely depends on various factors such as fertility of land, rainfall, irrigation, application of fertilizers, weather conditions, prices etc. In order to study the trend of increase or decrease with the previous year, crop-wise analysis has been made in the ensuing paragraphs. The production of food-grains during the year under report is 6146044 tonnes as against 4312023 tonnes in 03-04 recording an increase of 1834545 tonnes or 42.5 percent .The production of rice is dominant constituent, accounting for 82.4 percent of the total food-grains production in the state. Crop-wise production of food-grains is given in Table 10.3.

The production of cereals recorded an increase of 44.3 percent whereas the production of pulses recorded an increase of 7.6 percent when compared to the previous year.

Sources of Irrigation

Water is an important determinant factor of production of

crops in the agriculture sector. Intensive and extensive cultivation of land depend mainly on the availability of water. Medium and minor irrigation schemes are implemented in the state for augmenting the irrigation for agriculture. The various sources of irrigation are canals, tanks, tube wells, open wells and springs. The number of sources of irrigation is furnished in Table 10.4.

Table 10.3: Crop-wise Production of Foodgrains (in tones)

Crop		04-05		03-04	
		Production	%	Production	%
A. Cereals					
1.	Rice	5061622	82.46	3222776	74.74
2.	Jowar (Cholam)	252063	4.10	245933	5.70
3.	Bajra (Cumbu)	124300	2.02	172341	4.00
4.	Ragi	154085	2.51	176381	4.09
5.	Maize	294717	4.80	250992	5.82
6.	Korra	793	0.01	1091	0.03
7.	Varagu	10455	0.17	9876	0.23
8.	Samai	27866	0.45	27808	0.64
9.	Other cereals	3712	0.06	3769	0.09
Sub Total (A)		5929613	96.48	4110967	95.34
B. Pulses					
10.	Bengalgram	3942	0.06	4349	0.10
11.	Redgram	28979	0.47	27475	0.64
12.	Greengram	61760	1.00	53315	1.24
13.	Blackgram	82998	1.35	75920	1.76
14.	Horsegram	20110	0.33	18240	0.42
15.	Other pulses	18642	0.30	21757	0.50
Sub Total (B)		216431	3.52	201056	4.66
Total Foodgrains (A+B)		6146044	100.0	4312023	100.00

Area Irrigated

The net area irrigated by different sources during 04-05 was 2637198 ha as against 2147673 ha in 03-04 showing an increase of 489525 ha or 22.8 percent over the previous year.

The net area irrigated during 04-05 constituted 51.7 percent of the net area sown in the state. Villupuram district is at the top with 222021 ha Net area irrigated. However, the highest percentage of the net area irrigated to the net area sown was recorded in Thiruvarur District with 96.4 percent followed by Kancheepuram District with 86.9 percent whereas the lowest percentage was recorded in The Nilgiris district with 0.9 percent. Table 10.5 shows the net area irrigated by various sources compared with 03-04.

Table 10.4: Number of Sources for Irrigation (2004-2005)

Sources	**No.**
1. Canals	
a) Govt.	2,389
b) Private	2
2. Reservoirs	71
3. Tanks	
a) Ayacut above 40 ha.	7,933
b) Ayacut below 40 ha.	32,386
Total	40,319
4. Wells	
A) Tube Wells & Others	
a) Govt.	1,169
b) Private	2,50,684
Total	2,51,853
B) Open Wells	
a) Govt.	670
b) Private	16,07,854
Total	16,08,524
Total no. of wells (A+B)	18,60,377

Canals

Canals are the major source of irrigation in Tamil Nadu. As the system tanks get supply from a permanent storage like reservoirs, dams etc. the area irrigated by this source is classified under canals. During 04-05, Canal irrigation accounted for 28.6 percent of the net area irrigated by all sources in the state. The net area irrigated by canals including system tanks during the year under review was 753819 ha as against 449458 ha in 03-04 recording an increase of 304361 ha i.e. 67.7 percent. The net area irrigated by canals is the highest in Thiruvarur district with 141206 ha (18.7 percent) followed by Thanjavur district with 131037 ha (17.4 percent) and Nagapattinam district with 118174 ha (15.7 percent) of the total net area irrigated by canals in the State.

Table 10.5: River Water Sources of Irrigation

S.N	Name of the River	Village/Distt. (Origin)	State	Length (K.m.)	Catchments Area (Sq. Km)
1.	Varahandi	North Arcot	TN	94	3044
2.	Ponnaiyar	Kolar	KT	396	14130
3.	Vellar	Chithri Hills	TN	193	8558
4.	Vaigai	Madurai	TN	258	7031
5.	Pambar	Madurai	TN	125	3104
6.	Gundar	Madurai	TN	146	5647
7.	Vaippar	Tirunolvolli	TN	130	5288
8.	Tambraparni	Tirunolvolli	TN	130	5969
9.	Subarnarekha	Nagri/Ranchi	Bihar	395	19296
Total					248505

TN = Tamil Nadu; KT = Karnataka

Tanks (Non-system tanks only)

The non-system tanks which are fed partly from their independent catchments areas and partly from the diversion of river water and jungle streams depend fully on rain. The net area irrigated by non-system tanks during the year 04-05 is 4, 65,355 ha as against 384960 ha in 03-04 registering an increase of 80395 ha or 20.9 percent over previous year. The extent of area irrigated by tanks during the year is the highest in Pudhukottai district with 15.3 percent followed by Sivagangai district with 14.5 percent, Ramanathapuram district with 13.2 percent and Kancheepuram district with 11.6

percent.

Wells

Wells are a principal source of irrigation in Tamil Nadu. During the year under review, open wells and tube-wells continued to be the principal sources of irrigation. The net area irrigated by open wells and tube-wells together accounted for 1400394 ha (about 53.1 percent of the total net area irrigated) in 04-05 as against 1298627 ha in 03-04 showing an increase of 101767 ha i.e. 7.8 percent. The net area irrigated by wells during the year 04-05 was the highest in Villupuram district with 12.1 percent of the total net area irrigated by the wells in the state followed by 9.1 percent in Thiruvannamalai district and 7.9 percent in Coimbatore district.

Forest Resources

As per the State of Forests Report 2003 of Forest Survey of India, the forest cover of the state is 22,643 sq. km. which is 17.41 percent of the geographical area of the State.

The details are given in Table 10.6.

Table 10.6: Forest Cover of the State

Forest	Land (Sq. Km.)
Geographic Area	130,058
Very Dense Forest	2,440
Moderately Dense Forest	9,567
Open Forest	10,636
Total	22,643
Percent	17.41

The recorded forest area is 2.26 million ha which constitutes 17.40 percent of the land area of the state. Reserved forest comprises 86.12 percent, protected forest 11.17 percent and unclassed forest 2.71 percent. The Tamil Nadu Forest Department is the custodian of 22,865 sq. kms. of forest land in the State. Forests are nature's renewable resource, essential for environmental stability and food security. Forests help in regeneration of the soil by adding humus, which acts as a porous reservoir to retain water, helping to sustain agriculture in the hinterland. Forests protect water catchments areas, conserve soil, ameliorate climate and combat global warming and desertification. Forests need to be managed in a way that ensures their sustainable utilization, safeguarding the ecological process and genetic diversity essential for the maintenance of the resource base. At present, 17.58 percent of the geographical area is under forests as against 33 percent targeted under the National Forest Policy, 1988. Nearly half of the forest area is subjected to heavy degradation on account of biotic pressure.

Conclusion

The land use pattern in Tamil Nadu has been evolved as the result of the action and integration of various factors like the institutional framework and the structure of other resources. The present land use pattern and the trends during the recent years will help to suggest the scope for planned shifts in the patterns and increasing crop production.

References

1. http://www.tn.gov.in/crop/landusebattern/htm.
2. http://www.forest.tn.nic.in/htm.
3. http://www.tamilnadu/tn.htm.
4. http://www.resources.nic.in/htm.

11

Computerization and E-Governance in Dairy Co-operative Marketing Organizations

Samwel Kakuko Lopoyetum

Dairy co-operative marketing organizations are very significant in many developing countries such as India and Kenya. The Dairy co-operative marketing organizations play a significant role in ameliorating the rural economy in the provision of gainful employment opportunities and gainful income to the rural communities. The Dairy enterprise is one of the contributing sources of income for the grass root people, namely weaker sections and other socio-economic poor (backward) sections of the society. In fact, dairy farming system can prevent enmass migration of rural population to the urban areas by providing self-employment to unemployed youth, women, rural families etc. The effective and efficient measures taken by the Tamil Nadu Co-operative Milk Producer Federation and Milk Producers Unions has enabled approximately 7175 Village Milk Producers Co-operative Societies and Seventeen District Milk Producers Co-operative Unions in Tamil Nadu State to become more profitable, productive and more viable. There are elements of good working performance in spite of the cut throat competition in the dairy market under liberalization, privatization and globalization (LPG).

With the advent and emergence of LPG in almost all major areas of business, industry, trade and commercial undertakings, opportunities have greatly expanded and opened up the new Indian economic system to the entire globe, and exploitation of almost all natural resources is the order of the day. There is no second thought that liberalization, privatization and globalization process must continue, but utmost care must be taken into consideration while earnestly undertaking and accelerating the process of modernization, computerization, mechanizations and up-gradations of dairy co-operative marketing sector to save the entire co-operative sector from being swallowed by MNCs or foreign-based dairy business organizations.

The silent introduction of LPG policies, practices and procedures have agreeably impressed and affected the functions, objectives, management, administration, process, procedures, systems and operations of many co-operative business organizations including the dairy co-operative sector (treated as a third sector) and therefore, the need and necessity for dairy co-operative sector to accept the new economic environmental changes which is highly indispensable in the open market economic system. Even, the techno-economic changes have become very imperative and sin-quo-non-for the growth and development of the dairy cooperative sector in this era of LPG.

Consequently, almost all business organizations depend increasingly on Information Technology and Management Information Systems (MIS) which would enable the process of organizational environmental adaptation to the changing situation, resulting into making organizations more efficient and effective in their working operations and management style. The Information Systems and Communication or Information Technology Enabled Services (ITES) play a very crucial role for organizational adaptability which can eventually enhance the decision-making process in many dairy co-operative organizations. In this fast changing world, co-operative dairy organizations must be geared to accept the new emerging changes, challenges, threats and modify their operational strategies, technological strategies, inputs, communication, planning, organizational network, control, coordination and modern technologies as required by the dynamic global changes exacerbated by the new economic milieu. Moreover, in the new millennium, Information and Communications Technology (ICT) forms an important element or ingredients of business organizations in the era of globalized economic environment.

Components of Computer and Information Technology

There are distinct ways of analyzing the components of computer and information technology and can be identified for use in different sectors of the economy. They are:

- Technology and its application for different purposes in sectors such as private, co-operative, public, agriculture (primary sector), tertiary, secondary, industry and service sectors,
- Development of advanced equipments, machines and tools, and
- Development of software and hardware system etc.

Technology and its Application in Different Sectors of the Indian Economy

It consists of methods, system, procedures, practices and techniques for collection and transformation of data. Communication links between different geographical places and regions is of vital importance for the success of information technology based on feedback and control system. Today, high quality of computers is used for communication and different functional areas of organizations. The communication systems and networked environment of different sectors of the economy in present day are making the jobs of connecting different parts and geographical locations more easily and even becoming more economical and manageable in handling the information technology and communication system.

The development of equipment machines, tools and their peripherals are fast emerging in the computerization, mechanization and modernization process. It is even taking place at a faster speed and costs are reducing every time. It can take the dimensions of:

- Pattern recognition equipment such as, video cameras and scanners etc.
- Fiber optics and satellite etc.
- Other advance equipments, machine, tools and instruments.

The next component in information technology is the development and preparation of the software application. There is a relentlessly army of engineers, technicians and technologists (experts) for an effective and efficient use of technological innovations and equipment upgradation that have been brought in by consistent research and development in the field of information technology and communication development system. The hardware system is also being developed and upgraded from time to time.

Computer is the heart of all modern information technologies. It is an electronic digital device designed to accept input requirement. It can process it and display the output. Computer along with other input devices (keyboard, scanner etc.) and output devices (printer, plotter etc.) are referred to as hardware. At present, India's computer density is 0.7 for 1000 population against the world's average of 25 per 1000. The densities are bound to improve in the years to come.

Computerization and information technology have played significant role in almost all sectors of the economy including agriculture, industry, service, co-operative sector (all co-operative sectors including dairy co-operative business organizations), private, public, tertiary, primary, secondary, business and social sectors. This concept of computerization process and application is true for any developed and developing countries such as India and Kenya. In fact, human history has already witnessed the industrial revolution, economic growth and the agricultural revolution which includes Green Revolution and White Revolution (milk production system). Information technology and communication is emerging as the technological infrastructure revolution for global integration and rapid development of all sectors of the country.

In broad terms, computerization and information technology are dealing with all sub-sectors which includes generation, transmission and utilisation of information like telecommunication, computers, consumer electronics, media infrastructure etc. The information technology and computerization are also as capable of integrating all the development and innovation of different types of programmes/schemes. There has been a revolution of transformation in the economy from old to modern economy based on knowledge (knowledge management and knowledge workers). The development of information technology and communication and computerization process provides necessary information about the whole structure of the economy in general and necessary information about the market in particular. When the information technology is more efficient and effective it can directly and indirectly help to accelerate the optimum growth of the economy and promote development and growth in all sectors.

It is felt, that information and communication technology and computerization today has become an integral part of the development process of the economy, if only it is geographically and politically (geo-politically) fairly distributed in the country to include the coverage of co-operative sector and the entire rural economy. Planners can avoid digital divide in the changing business environment.

Relevance of Computerization of DCBOs

In order to attain significant rural development and enhanced rural prosperity, eminent scientists such as Prof. M. Swaminathan advocated for bridging the existing gap between the digital divide among the affluent communities and non-affluent community. In fact, Prof. M. Swaminathan, Chairman,

National Commission on Farmers, Founder, M.S. Foundation and Research Centre, the architect of India's Green Revolution, was always concerned that all the benefits of science, technology growth and development do not reach the poor. It has not reached the ultra poor even. He was also greatly concerned about the tremendous differences caused by the new technologies, especially information and communications technology, which could bridge the gap existing between the rich and the poor. He wanted to use the very same information and communications technology to bring about greater opportunities for the poor and to usher in an era of job-lead growth. Thus, the ultimate aim is to enhance the growth and development of the rural economy and enhance socio-economic development of rural communities and improve their livelihood management system in the 21st Century. The following are the merits.

The computer was initially invented as a high-speed calculator. This has largely led to enormous progress and many scientific projects which were previously impossible to undertake without the use of computers are now enjoying the application and benefits.

In fact, electrical pulses travel at an incredible speed and operations. The units of speed are measured in the microsecond (millionths). The nano second (thousand-millionths) and later even the microsecond (million-millionths). It is even interesting to note that powerful computer is capable of adding together two 18 digit numbers in 300 and 400 nanosecond. These benefits and significance can be useful and applied in the development of the service sector, dairy co-operative business organizations, agriculture, industry, and rural development organizations in the 21st Century.

- **Storage of Information and Data:** The speed with which computers can process large volume of information has led to the generation of new data and information on a vast scale. In other words, the computer has compounded the information explosion or electronic explosion. IT can store data and information for use in many sectors of the economy including the dairy co-operative business organizations.

- **Accuracy of Computers:** The computers accuracy is consistently becoming higher and higher. Almost, without exception, the errors in computing are due to human rather than to technological weaknesses, i.e., imprecise thinking by the programmer, inaccurate data, or poorly designed systems can contribute to inaccuracy of computers. Accuracy can be useful in taking appropriate decisions and relevant data and information in the service sector, dairy co-operative business organizations and rural development organizations.
- **Versatility:** Computers are capable of performing almost any task, provided that the task can be reduced to a series of logical steps and operations.
- **Automation** A computer is much more than an adding machine or calculator which requires human operators to press the necessary keys for the operations to be performed. Once a program is in the computers memory, the individual instructions works until the last command which says stop program execution is administered, then it can stop likewise.
- **Diligence** Being a machine, a computer does not suffer from the human traits of tiredness and lack of concentration etc. If 2 million calculations have to be performed, it will perform the 2 millionth calculations with exactly the same accuracy and speed as in the first instance.

Classification and Application of Computers

Large number of organizations requires the most powerful computers, while individuals may be satisfied with a personal computer (PC). The computers are classified according to the size and memory capacity. A few of the classification and application are as follows:

1. **Personal Computer:** A personal computer or a microcomputer is the smallest general purpose computer system, which can execute programs to process a variety of instructions. It has all the functional elements found in a larger system. These computers usually have an 8, 16 or a 32 bit micro processor. The cheapest variety is one with an 8-bit microprocessor. All 8-bit micros are build around a few popular microprocessors like the z80, 8080 etc. By an 8-bit microprocessor it is understood that it can process 8-bits or 1 byte of data at a time. These microprocessors have a 16-line address bus and hence they can identify a maximum of 2.16 i.e. 64,536 or 64k address locations etc.
2. **Mini Computers:** A minicomputer system is a small general-purpose computer varying in size, from a desktop model to a unit size of a four-drawer filing cabinet. It is quite obvious that, there is tremendous amount of similarity between the more powerful micros and the lower end minis. The same situation exists on the other end where the lower priced mainframes are almost similar to the higher priced minis. Today's typical mini will surpass the PC in storage capacity, speed of arithmetic operations and ability to support a variety of peripherals. Minis are usually with the 16-bit microprocessor but now-a-days they are available only with 32 bit.
3. **Mainframe Computers:** A whole series of mainframe models, ranging in size from small to very large, are typically lumped together under a family designation by mainframe manufacturers. Several microprocessors are used in place of the single microprocessor used in micro and minicomputer system. The result is that these systems can process data much faster. They also have the flexibility to operate automatically from 2 to 8 bytes in the same unit of time. The mainframe vendors also have much large libraries of application programmes.
4. **Super Computers:** The supercomputers are the most

powerful and expensive computers available. Only a few of these computer monsters are made each year because only few people need it. They are designed to perform scientific application and hence the computational speed is very important. To maximize the speed each address location holds 64 bits of information. The time required to execute a single operation may be as slow as nanoseconds.

5. **Software:** Software is the means of communicating with the computers. The set of instructions written in a logical sequence are referred as software. They are broadly grouped under two categories: System Software and Application System. System software is the base level software needed to run any application. It has to be present for the application software to function. This software determines which program gets priority in the use of the various resources, such as disk or printer or memory. It translates the application software into a form understandable by the computer.
6. **Application of Software:** Application software is a software designed to specifically perform a particular user function or application. This software is for the user to operate, the computer to do the processing and give the results needed. In the world of computers, change is permanent than anything else. The change happens like ripples in a tank of water. First improvements are incorporated in hardware followed by system software and finally application software.
7. The computer users are bound to increase as the cost of PC falls at the rate of 40 percent in a year. The prices of computers, PC and taxation of the equipment machines, tools, and peripherals have declined tremendously as recommended by the Ministry of Finance, Government of India.
8. **Internet Facilities:** Internet is a collection of tens of thousands of networks of computers across the globe. It is a vast, worldwide system consisting of people,

information, and computers. Internet is as large and complex as to be well beyond the comprehension of a single human being. Nearly 100 million users have access to Internet and will double within the shortest period. Even perhaps, the best way to understand the organization and importance of the Internet is to compare it to the other great worldwide communication systems like the postal system and the telephone system (communication system). Co-operative business organizations can utilize the emerging and modern Internet facilities in a big way, to compete with their competitors or rivals in the new economic environment.

9. **E-mail:**By using E-mail, one can send a message from one computer in the net to any other computer across the Globe. It requires one to run the e-mail programmed, type the message and the receivers address and click a button. The mail popup on the receivers screen a few moments later. If there are no hardware problems, the reply can also be obtained within few minutes-practically for the cost of a local phone call. As per several surveys, the number of host computers in Internet crossed the 10-million mark and the annual growth rate is 85 percent per annum.
10. **Newsgroups:** These are electronic boards with a worldwide fan club, each devoted to some special interest. One can get information, ask questions, canvass opinions, pass on bit of news or get advice from this group-all related to that particular subject. Other users in the group are supposed to read what one has posted and within a few minutes there will be a new posting in response to the earlier query-may be an answer or an explanation. There are newsgroups already for conceivable ares of human interest, from agriculture, industry, service to hobbies etc.
11. **World Wide Web (WWW):** The web is an ambitious client/serve system. This means that you can operate a client (which runs on your computer) to contact a server program (which runs on another computer, some where in

the internet). The web offers a simple interface to the vast resources of the Internet. Getting started is easy and just about anyone can start enjoying the web with little preparation.

Moreover, the benefits of Information Technology (IT) have to reach the common man and they should have access to these services even if they are in the remotest part of the country. The convergence of different forms of Information technology services viz., television, radio, telephone, internet in a single unified environment; is now increasingly possible to have access to information technology services etc.

In the new millennium, there are huge benefits and contributions of the new technological advancement of the equipments, computerization, tools, processing system, materials, machines and complex machineries in the production and processing of milk and milk products. In fact, co-operative dairy marketing business organizations have now made use of the latest machines, instruments and equipments such as Electronic Milks Tester (EMT), Microprocessor based milk collection system, Milk Weighing System (MWS), Data Processor (DP), PC based milk collective system, Automatic Milk Collection Station (AMCS) etc.

E-Governance of DCBOs

E-co-operative system (e-governance) is the application of information technology and communication process to transact all the business processes of co-operative organizations effectively. Efficient functioning, responsiveness, prompt dissemination of information, prompt delivery of quality services, improved productivity and profitability by using the latest skills and knowledge of information technology and electronic revolution are some of the prerequisites.

E-governance of co-operative dairy business organizations is simply the ability of the dairy cooperative organizations to provide all access to services and products through electronic and mechanized production and marketing system. Even, dairy co-operative business organizations can provide on-line services because of very huge opportunities due to revolution in the information technology and communication systems, which have made everything possible in the business and e-business world. Even electronic purchase (e-purchase), payment (e-payment), customer transactions, e-CRM, business activities or operations of all types etc. are made electronically with the complete use of the latest methods of information technology and communication. Moreover, through e-governance, e-business and e-commerce, and effective process of mechanization and computerization, the working and performance of dairy co-operative organizations can be improved in the 21st Century.

Relevance of Computerization of DCMOs

Relevance of computerization and e-governance in dairy co-operative marketing business organizations is extremely beyond imagination. E-governance of co-operative dairy organizations is simply the ability of the dairy co-operative marketing organizations to provide all access to services and products through electronic and mechanized production and marketing system. The dairy co-operative organizations can provide on-line services because of dynamic revolution (electronic revolution) in Information Technology and Communication (ITC) system, which has made everything possible (transactions) in the business and e-business world. Even dairy purchases (e-purchase), milk payment (e-payment), customers transactions, business activities or operations of all types etc., are made electronically with the complete usage of the latest methods of Information and Communications Technology (ICT).

Significance of District Milk Producers Co-operative Unions in the Changing Dairy Business Environment

A very large number of dairy farmers are directly involved in the business affairs of Dairy Co-operatives Societies (DCS) which would eventually double in the 21st century with the emergence of new generation dairy farmers and emergence of progressive self-help groups. There would be a reduced dependency on imported dairy products and commodities due to high domestic production of milk. There are huge prospects of sustainable growth and generation of adequate rural employment through dairy co-operative business organizations in the 21st century. Transfer of income from urban to rural areas and equitable distribution of income would be streamlined in the years to come.

Already, there was large spin off of rural economic development through Operation Flood programmes of NDDB i.e., OF I, OF II, OF III and other new dairy development programmes such as vision 2010 of NDDB. By creating technological consciousness and awareness among 75 percent of 'OF' beneficiaries who are landless, marginal and small farmers, there is strong possibility of dairy sector moving forward in a big way in the era of economic liberalization, privatization and globalization.

In fact, the Vision 2010 plan of NDDB is geared towards production of milk and milk products in different milk shed areas and dairy regions of the country in large quantities in the forthcoming years. They must eventually aim to develop almost all the district milk producer co-operative unions and ensure higher socio-economic benefits to the dairy/milk producers in the rural areas. Notwithstanding modernization, computerization, mechanization and advancement of science and technology in the 21st century, there will be tremendous social empowerment and emancipation of rural poor leading to significant improvement in their socio-eco-political environment.

Women particularly, would be the largest beneficiaries of the 21st century sustainable dairy co-operative sector. Presently, the animal husbandry and dairying industry contributes 8-10 percent to the GDP of the country, which would be enhanced in the ensuing decades. All the major dairy business activities will be mainly located at the dairy union level.

The District Milk Producers Co-operative Unions must diversify their dairy business functioning, management and administration. They must improve the availability of cross-breed animals, recommend high yielding animals and must intensify and extensity artificial insemination (AI) services, improve the provision of concentrate feeds to farmers, improve fodder production and milk collection operations, improve milk distribution, venture into production of different milk products, improve organization of most of the PMPCS. The dairy union must provide adequate technical inputs from time to time for achieving the above goals. The usual operations of the union is collecting milk from its members for processing into various products and also selling unprocessed milk to meet the share of milk demand in most parts of the Districts.

By collecting milk from PMPCS, they are directly boosting their business performance and enhancing profitable working performance of the PMPCS. They must initiate programmes to enhance production of milk at the District level through dairy extension services, promotion of dairy animals breeding programmes etc. They must work in tandem with the support system and dairy/milk producers themselves in collaboration with the departments of animal husbandry, cooperation, and other related departments.

Moreover, the most important way of harnessing the livestock wealth in most of the rural and urban areas can be through an effective and efficient way of managing and administering the business operations of the cooperative dairy sector, private sector, self-help and voluntary (business) organizations. It would generate adequate income and gainful employment to the enormous number of rural dairy farmers. One advantage is that, the cattle feed and fodder sources are mainly based on the crop residues, open grazing system, collection of fodder from various places and obtaining concentrate feeds from the PMPCS stores.

In the changing business environment, the milk producers union must adopt the instruments of the backward and forward linkages and fill gaps between demand and supply of milk in the Districts. It must sensitize rural dairy farmers on the cost of production of milk, seasonal fluctuations, enhancing their strategies on the quantity and quality of the milk produced. The District Milk Producers Co-operative Unions must arrange for prompt payment i.e., weekly or fortnightly as desired by the PMPCS members and ensure lower rates of concentrated cattle feed, than the prices offered by private supplies/businessmen (Anand Dairy pattern advocate for this approach). In the procurement and marketing operations of milk producers cooperative union, the critical areas which always requires greater attention are milk quality control, packaging of milk and milk products, branding, advertisement, milk pricing, milk payment and milk processing.

The union must ensure that all these factors are managed and controlled to ensure higher profitability and ensure sustainability in the era of cut throat competition prevailing in the open market economic system. However, privatization scenario and globalization environment are posing greater challenges to the growth and development of dairy co-operative marketing organizations.

Insights into Madurai District Co-operative Milk Producers Union Limited

Historical Origin and Development of Dairy Union:

The Madurai Milk Project was implemented by the Government in the year 1967 with the financial assistance from UNICEF. Madurai Dairy Co-operative Union was equipped with manufacturing plant, equipments and machineries provided by the UNICEF and New Zealand Freedom from Hunger Campaign Committee, New Zealand. In order to feed milk to Madurai Dairy, the chilling centres located at Virudhunagar and Srivilliputhur were constructed during 1976. It was processing 50,000 litres of milk per day from 01.10.1974. Tamil Nadu Dairy Development Corporation Limited had procured milk from Madurai, Theni, Dindigul, Virudhunagar and parts of Coimbatore and Dharapuram milk co-operative societies.

The formation of Madurai District Cooperative Milk Producers Union Limited started from 01.10.1983. During that time, milk procurement was restricted to Theni and Madurai Districts only. It was registered under the MMPO registration Certificate No.127/R-MMPO/94 dated 9.2.1994 for a capacity of 2.0 lakh KPD (renewed for a period of 5 years). This registration was issued by the Government of India. The outlay for Operation Flood II was Rs. 126 lakh and Operation Flood III was Rs. 79 lakh. The capacity was: liquid milk processing – 1.5 lakh litres per day; powder production – 10.0 M.T. per day; butter production - 10.0 M.T. per day and ghee production – 0.5 M.T. per day. The shield was awarded to this Union for the best functioning district union both at State and Districts level.

Main Objective of Dairy Co-operative Union: The main objective is to uplift the socio-economic status of rural producers by ensuring reasonable price for their products and satisfy the demands of the consuming public in Madurai region at a reasonable price.

Milk Procurement Teams: The Dairy Union is procuring milk from their own societies and from other unions such as composite Madurai District, Tirunelveli, composite Ramnad District and Kanyakumari District. The team identifies the milk potential areas, organise milk co-operative societies, form viable milk collection routes, enlighten the milk producers to use suitable cattle feed, raise fodder in their land and educate them on the economic functions of milk co-operatives. There are 9 mobile veterinary routes which operate in this Union and 40-50 Societies are covered per route under the health cover.

The following health coverage schemes are extended to the animals through the dairy co-operative societies. 1. De worming of calves, 2. Treatment of sick animals, 3. Sterility camps, 4. AI and Pregnancy verification 5. Insurance

Coverage, 6. Immunization of animals against contagious diseases, 7. Cattle feed supply, 8. Fodder development, 9. Education and training, 10. Emergency service offered in various milk collections routes: The routes are: (a) Usilampatti (b) Melur and (c) Theni. The service charges amount to: Rs. 25 per case and average cases are 200/month. Through 576 functional societies, the Union is procuring 1,10,000 litres of milk everyday. By operating 21 milk collection routes, the milk produced by the primaries are transported to the Madurai Main Dairy and Theni Chilling Centre.

The Union is presently marketing around 90,000 litres of milk average per day to the consumers through 373 milk booths, 20 milk parlours and 45 retail outlets in Madurai City. A separate sales promotion wing namely "Consumer Oriented Milk Sales Promotion Wing" was formed to meet out milk requirements of hoteliers, retailers and special order booking parties at their doorsteps. The Union is also selling milk in Theni by engaging retailers and positioning suitable parlours in Theni and Cumbum areas. They need to expand their selling and marketing outlets in the future.

Calf Rearing Scheme: The Heifer Calf Rearing Scheme was implemented in 1977. The calves were included under the financial assistance from Government of India and Government of Tamil Nadu during the period 1988 to 1992. From 1992-93 the Calf Rearing Scheme was abolished. However, the Union Board resolved to implement the Scheme from Union Surplus Funds at a total cost of Rs. 4 lakh (2000 calves) under the Calf Rearing Scheme during the year 1993-94.

Formation of Strong and Viable District Milk Unions and Primary Milk Co-operative Society: In the era of LPG, there is an urgent need to encourage the new formation of strong and viable District Milk Producers Co-operative Unions and Primary Dairy Co-operative Societies. There is an urgent need for completely strengthening and advocating for frequent improvement of the working conditions and enhancement of the profitability, productivity, viability, sustainability, and business results of District Milk Producers Cooperative Unions and Primary Milk Producers Co-operative Society to feed the increasing population with adequate milk and milk products. There is an urgent need for inculcating the collective efficiency in all the District Milk Producers Co-operative Unions and Primary Milk Producers Co-operative Organizations in the present changing business environment. The District Milk Producers Co-operative Unions must be computerized and e-governed from time to time.

Dairy sector is known for granting permanent income and generating gainful employment opportunities to the weaker sections of the community, small and marginal farmers and the landless agricultural labourers. Dairy co-operative organizations encourage members to supply milk through co-operative societies (not vendors or private functionaries) and they can mobilize necessary support from rural dairy farmers and farmers associations. They must be motivated and psychological tuned to supplying milk to their own dairy co-operative societies and not to private dairy traders. Ultimately the PMPCS can certainly improve the overall socio-economic conditions of the dairy members in the long run. It is an effective agent of socio-economic transformation of both rural and semi-urban areas of developing countries such as Kenya and India.

Major Objectives of the Research Study

The major objectives of the research study are:

- To study the inherent factors responsible for embracing computerization and mechanization of dairy co-operative marketing organizations,
- To examine and assess the benefits of computerization and mechanization of dairy co-operative marketing organizations,
- To know and elicit the response of dairy co-operative officials on the relevance of e-governance and

mechanization (computerization) process,

- To identify the core areas where computerization and mechanization are having more effect and impact on dairy co-operative marketing organizations,
- To offer suggestions and policy alternatives for the development of computerization and mechanization of dairy co-operative marketing organizations.

Methodology and Tools of Analysis: The present research study is based on case study method and descriptive analysis and adoption of Focus Group Discussions (FGD's) with employees of dairy co-operatives marketing organizations. The researcher interacted directly with the respondents (employees) and total of 120 employees were interviewed and suitable statistical tools were used to derive inferences.

Basis of Selection of Respondents (Employees): The respondent interviewed constitutes higher (senior) level, middle level and lower level management cadres of Madurai District Milk Producers Co-operative Union.

Type of Data Used: This research study relied predominantly on primary data collected from the dairy co-operative staff (employees) of Madurai District Milk Producers Union. It makes use of focus group discussion (FGD) outcomes. Besides, the use of secondary sources of data and information obtained from authenticated sources and organizations is also used.

Major Findings

1. **Benefits of computerization and mechanization:** The study reveals that, all the respondents (100 percent) were most favourable to computerization process and mechanization of dairy co-operative business organizations, especially those in the rural based areas. This would help to empower the weaker sections (backward) of the community and the dairy farmers to overcome their operational problems (difficulties) through

computerization, modernization and mechanization of their dairy co-operative societies.

2. **Provide higher productivity and profitability and better Customer Relationship Management (CRM):** The study reveals that, cent percent of the respondents were most favourable. This is mainly due to the ability of computers to be used to transact at lower cost of management and administration, at the same time improve the overall effectiveness and efficiency of transactions and operations of the dairy co-operative organizations.
3. **Customer-related problems:** The application of computers and use of electronic equipment (mechanizations) will help to solve customer related problems speedily and accurately. The study reveals that, all the respondents were most favourable to computerization and mechanization of the dairy co-operative services such as input services and output services. By fully undertaking computerization and mechanization, the customer's problems are dealt with accurately, resolved in the most effective and efficient manner.
4. **Training programmes on computerization and e-governance:** The study reveals that, 80 percent were favourable and 20 percent were neutral. This is due to the fact that, it is significant to impart skills and knowledge to the employees on e-governance and computerization of dairy co-operative organizations.
5. **Urgency of computerization and mechanization:** Further, the study reveals that, 85 percent were most favourable for urgent computerization, modernization and mechanization of dairy co-operative organizations and its primary milk producers co-operative societies. There is an imperative need for complete modernization and computerization of dairy co-operatives in the new centenary and millennium i.e. computerize or perish can be adopted as slogan for co-operative dairy marketing

business organizations in order to undertake effective and efficient global dairy business activities.

6. **Insecurity of dairy co-operative employees and staff members:** It becomes interesting that, all the respondents (100 percent) were unfavourable to the installation of computerization and mechanization process. They stated that, computerization and mechanization would render the dairy co-operative employees insecure at all levels of management and administration. This is mainly because there would be no further employment opportunities being created for the prospective job seekers in the future, if the computerization and mechanization process were installed in the dairy cooperative organizations.
7. **Stress and strains of work:** Stress is the inner pressure caused by factors such as endogenous, exogenous, and interaction of endogenous and exogenous factors. The workers can undergo mental and emotional strain leading to psychosomatic disorders. It is interesting to find out that, 80 percent of the respondents were neutral on account of reduction of stress and strains caused by installation of MIS and mechanization (computerization) process.
8. **Organizational and structural changes:** The study reveals that, all the respondents were most favourable for organizational changes in dairy co-operative union and primary milk producer's cooperative societies. Whereas, 70 percent of the respondents were unfavourable for mandatory structural changes and mechanization of dairy co-operative organizations
9. **The impact on the dairy co-operative working procedures and administration:** The study reveals that, 65 percent of respondents were most favourable on the impact on reduction of the procedures and efficient administration of dairy marketing cooperative organization. This is mainly due to the merits and benefits derived from e-governance and computerization cum mechanization process on the overall operations of the

dairy marketing business organizations in the new millennium.

10. **Effect on the overall utilization of resources and management in dairy co-operative marketing organizations:** It is essential to obtain better outcome and benefits from all transactions carried out through computerization and mechanization of dairy co-operative organizations. In order to increase profitability, productivity and efficiency in management and administration of dairy cooperative organizations, it is imperative to pursue for the implementation of computerization process, e-governance and mechanization in dairy co-operative marketing business organization. In this context, the study reveals that, 75 percent of the respondents were most favourable on the computerization process and e-governance facilitating efficient and effective overall management of co-operative dairying organizations. In the changing economic environment dairy co-operative marketing business organizations can become popular by undertaking computerization and e-governance process. The success, prosperity, and posterity of these institutions to a larger extend depends on computerization, ICT, and e-governance of their functions, operations, transactions, management, administration and total business operations of dairy co-operative union. They are likely to gain competitive advantage and upper edge over their competitors, if only they computerize. They can offer improved services, become more profitable, productive and sustainable in the new economic environment.

Strategies for Promoting Development of Computerization Process and E-governance of Dairy Cooperative Marketing Business Organizations

The following are generic strategies to enhance growth and development of computerization process, e-governance and information and communications technology (ICT) in co-operative marketing business organizations in the ever changing dairy business environment.

1. **Careful Planning and Control:** While formulating the information technology and communication plan, computerization process, e-governance etc., the co-operative authority and policy planners should give more attention to implementation of the technology effectively in dairy co-operative marketing business organizations. Every plan of action on computerization process, e-governance, information and communications technology (ICT), should be made to ensure proper and suitable implementation of information technology and communication in the spirit they are made to attain higher growth rate in the long run.
2. **Need for Effective and Efficient Coordination, Co-operation and Communication Process:** To ensure effective and efficient co-ordination and co-operation among dairy co-operative marketing business societies, they must work in close tandem and contact with other co-operative business organizations like credit, consumer and processing co-operatives etc. This can be established to speed up the computerization and e-governance of the co-operative sector and should pave way for the growth of information and communication system horizontally and vertical co-ordination among various sub-systems of the co-operative business network is highly essential.
3. **Utilization of Technical Consultancy and Promotion Services:** In the era of electronic and technological revolution, no business can survive without the use of modern technology i.e. computerization process, e-governance process, information and communications technology (ICT). Therefore, it is a major responsibility of the Government and technical consultancy agencies concerned with the co-operative business organizations to serve as reliable grass-root co-operative consultancy

agency for the co-operative production societies, co-operative dairy marketing organizations, co-operative service and processing. They must give technical guidance in computerization and e-governance matters in various fields, aspects, operations and functions of dairy co-operative marketing business organizations.

4. **Modernizations, Up-Gradation, Computerization of Dairy Processing and Marketing Operations of Dairy Co-operatives Business Organizations:** Every dairy co-operative business society must have complete computerized information system. They can modernize, up-grade and computerize their dairy business operations so that it can record all the information about the dairy society as well as the members and customers transactions. By doing this, it can easily locate the problem in a few seconds and save their time through solving their problem promptly. Through computerization, e-governance and improved information system it will be easier for the dairy co-operative society to educate its members regarding the changing trends in the market and educate them about the latest trend in the technology and communication system.
5. **Need for Effective and Efficient Education Programmes on Computerizations, E-governance and Modernization Process:** Co-operative officials, staff, members, dairy farmers should be educated on modernization of the dairy sector in almost all its operations in various ways and procedures of handling dairy marketing system, agricultural marketing and processing through appropriate modernization process, e-governance dairy and proper computerization process of co-operative processing, operations, and transaction of dairy marketing societies. They can be educated through village knowledge centres, information centres, adult education centres, agricultural research centres, and co-operative institutions from time to time.

- There is an imperative need for setting up of dairy village co-operative knowledge and information kiosks/shops that would enable rural dairy families to access a basket of modern information and communication technologies. Motivation and training of educated youth, especially women, in rural areas in operating computers, information and knowledge centres in the rural economy is mandatory and highly essential in the new economic environment.

- There is an imperative need for training of the rural youth in the organization and maintenance of the dairy co-operative knowledge system that can generate adequate information technology and communication. Moreover, locally relevant information from generic information and data can be plainly used in the operation of the information and documentation centre through dairy cooperative business organizations.
- The dairy cooperative information and communication centre can be set up urgently to undertake the maintenance, updating and dissemination of information on entitlements to rural dairy co-operative marketing societies by using an appropriate blend of modern (computerization) and existing channels of communication. This must be made essential, part and parcel and necessary components in the dairy co-operative marketing information and documentation system to cope up with the changing business environment.
- Before installation of the dairy co-operative information and documentation centre, it is essential to conduct impact assessment analysis based on the organization of surveys, Participatory Rural Appraisal (PRA), Rapid Rural Appraisal (RRA), Participatory Action Research (PAR), Participatory Learning and Action (PLA), Focus Group Discussion (FGD) and other appropriate methods of data and information gathering for the growth and development of the dairy co-operatives, computerization, e-governance and documentation of the entire dairy co-operative marketing business organizations.

- There is an imperative need for building up of an appropriate model in dairy information dissemination, documentation facilities and exchange units in rural dairy co-operative societies that use advanced information, computerization, e-governance and communication technologies in the rural economy.
- The dairy co-operative information and documentation centre can start by elaborately conducting action oriented research on computerization, e-governance and the formation of multi-sectoral partnerships (Private-Public/Government/NGOs) with rural communities to form self supporting, self sufficient and sustainable model of information and communication technologies for rural areas in the changing economic scenario.
- Development of dairy computerization process, information and communication technology based applications for rural dairy co-operatives, especially online dairy system and online dairy industry, dairy services outline, dairy co-operative education and training to assess their potential to contribute to the sustenance of rural dairy co-operative information and communication technology programme in the context of the changing economic environment and globalization scenario.
- There is an imperative need for organization of dairy co-operative workshops, training and development programmes and other interactions such as on-site consultations with dairy co-operative policy makers to sensitize them to critical issues in the use of co-operative information and communication technologies, computerization process, e-governance and modernization of dairy co-operative marketing business organizations.
- There are huge benefits accrued from computerization, e-governance, information and communication technology such as gaining fully from the exchange of research results with dairy ICT research stations and telecenter programmes in different part of the country (co-

operatively undertaken) to arrive at a more comprehensive picture of impact assessment, productivity, profitability, economic viability, and sustainability issues and possible internationalization of co-operative project implementation and execution in the changing global business scenario and eventually enter into 21st Century with posterity and prosperity.

- Setting up of the dairy co-operative marketing information and communication technology centres will help in the development of a set of parameters to make a choice of technical and market access of dairy technology and rural technology for powering the real access devices to the milk producers and the entire rural economy in the long-run.
- There is need for the development of models of computerization, e-governance, co-operative eco-jobs and eco-co-operative entrepreneurship among rural milk producers and develop rural economy in the 21st century.
- Through co-operative information and communication technology centre/kiosk, it would be possible to link the entire co-operative sector and promote dairy co-operative entrepreneurship to enhance sustainable development of the rural economy in the 21st century.
- The dairy co-operative entrepreneurship and computerization can be incorporated all aspects of technical, managerial and marketing capacities of the rural milk producers to mobilize themselves as dairy co-operative entrepreneurs producing high quality milk and milk products and services for sustainable dairy development with the help of the co-operative information and communication technology in the 21st century.
- Through dairy co-operative information and communication technology, the concept of co-operative bio-village and dairy entrepreneurs can be promoted. It is an approach which strengthens the capacity of the rural milk producers to blend sustainable natural management

with livelihood security through economically feasible, socially acceptable, economically viable and gender-sensitive interventions under the co-operative umbrella.

- By setting up the dairy co-operative information and communication technology kiosks/centres and complete computerization of dairy co-operative marketing organizations, a people-centered approach can facilitate the socially and economically disadvantaged group of the milk producers vis-à-vis class, caste and gender etc.
- Through the dairy co-operative information and communication technology kiosks/centres the support of funding agencies can be sourced and used by the milk producers as a leverage to gain access to the commercial bank credits and government resources can be solicited from time to time.
- There is an imperative need for evolving grass-root institutions such as the primary milk producers co-operative societies to take up development initiatives in computerization, e-governance, and setting up an effective and efficient dairy information and communication kiosks/centres in the rural areas.
- Through the process of computerization, e-governance and dairy co-operative information and communication technology, it will be possible to build social capital and linking it with economic capital in the long-run and enhancing the ability of rural milk producers in defining the path of sustainable development in the 21st century.
- Communication information empowerment through dairy co-operative village information and knowledge centres/hubs would lead to the information empowerment through computer-aided and Intra-net connecting dairy villagers for horizontal transfer of knowledge in the rural economy in the 21st century.
- The dairy co-operative information and communication technology societies can be treated as local-specific information empowerment centres for rural dairy

communities leading to lesser social and economic transaction cost.

- It can be initiated as a hub and communication model and each dairy co-operative information and communication technology centre/kiosks can be managed by progressive SHGs or new generation dairy co-operative or milk producer co-operatives in their respective villages in the long-run.
- The dairy co-operative information and communication technology system can convert and disseminate generic data and information into local specific information documentation to assist the rural dairy co-operative societies in decision-making process.
- The dairy co-operative information and communication technology centres/kiosks can communicate information and data every day on various aspects of dairy development to all interested persons and parties in the rural areas.
- The dairy co-operative information and communication technology system can also be used to train villagers in areas like sustainable agriculture, functional literacy and computer literacy programmes and also to train semi literate villagers and other villagers, to manage the adult literacy programme by using computers. This can also be done by setting up a website which provides, all information and data and the website can be managed through co-operative approach/sector.

Conclusion

This research paper has elucidated and analyzed all the inherent factors and also provided highly relevant major recommendations, policy options and appropriate suggestions to enhance efficiency, effectiveness, profitability and productivity through e-governance and computerization of the dairy co-operative business organizations in the new economic environment.

References

Anjani Kumar, (October-1998) "Milk production function and factor demand in milk production", The Bihar Journal of Agricultural Marketing, Vol. 4, No. 4.

Gupta. S.K. (Nov.1999) "Dairy in next millennium-The challenges" Yojana Vol. 43 (11).

Deepak Shah and Asok, (1996) "Impact of dairy co-operative on marketing pattern of milk – A case study", Indian Co-operative Review, No. 33 (3).

Hage N.C., (2001) "WTO challenges for Indian dairy farmers", Yojana.

Iyer L.L., (1973) "Fifty years of Dairy Development in India tasks ahead", Indian Dairy man, vol. 25, No. 12.

Katar Singh of Mukunda Dar., (1982) "Impact of Operation I at the village level – research report", Anand IRMA.

Koli. P.A (July 2000) "Can milk co-operatives sustain in the 21st century", Indian Cooperative Review Vol. XXXVIII, No. I.

Kulandaisamy, V., (1982), "Co-operative Dairying in India", Rainbow Publication, Coimbatore.

Mohinder Singh and Anil Kumar, (1996) "Technological interventions in rural dairy development: a study", Kurukshetra.

Nath Abbay et.al., (1998) "Milk procurement a challenge for personnel of milk union", Rural India Vol. 61, No. 6.

Samwel Kakuko Lopoyetum, (2004) "Good Governance through E-Governance" National seminar conducted by Department of Political Science and Administration, GRU.

Samwel Kakuko Lopoyetum, (July 1999) "Computer Application in the Management Administration of Cooperatives" Tamil Nadu Journal of Cooperation, Vol. 91, No.4, pp. 51-58.

Samwel Kakuko Lopoyetum, (October-December 1999) "Indo-Kenya Dairy Cooperative System: Scope for Dairy Cooperative Development and Replicability Strategy of Indian Model in Kenya" Cooperative Perspective Journal of Cooperative Management (Quarterly issue), Vol.34, No.3, pp. 21-27.

Samwel Kakuko Lopoyetum, etal, (2001), "Members perception of the social impact of Dairy Cooperative Societies in Dindigul District: A Survey", Indian Cooperative Review, 39 (1).

Selvaraj. V. (September 2002), "Marketing of Dairy Products through E-Commerce", Kissan World, Vol.29.

Shiyani R.L., (1996) "An economic enquiry into the impact of Dairy Cooperatives on milk production", Indian journal of agricultural economics No. 51 (3).

Shrode, (1999) "Dairy in India", Rural India, Vol. 62, No. 9.

Singh Mohar et.al., (1998) "Constraints faced by the dairy seeking information regarding improved dairy farming practices", Rural India, Vol. 61, pp. 9-10.

Srinivasan Gowda, M.V., "Cooperatives in the new economic scenario: agenda for liberalization", Southern Economist, p. 21.

Subburaj. B., Ravichandran K., and Samwel Kakuko Lopoyetum, (February 2003), "Appropriateness of Management Strategies, Are they myth to Cooperatives", The Co-operator, Vo.40, No.8, February 2003.

Appendix

Appendix

National Policy for Farmers, 2007

The National Commission on Farmers (NCF) [chairman: M.S. Swaminathan] submitted its final report in October 2006. The NCF prepared, along with its final report, a draft National Policy for Farmers incorporating its main recommendations. Based on the draft prepared by the NCF and after consultations with State Governments and the Central Ministries concerned, Government of India approved the National Policy for Farmers, 2007. The objective is, inter alia, to improve the economic viability of farming through substantially improving net income of farmers. There is emphasis on increased productivity, profitability, institutional support, and improvement of land, water and support services apart from provisions of appropriate technology, price policy, risk mitigation measures and so on.

Need for Policy Re-orientation

The importance of agriculture in the socio-economic fabric of India can be realised from the fact that the livelihood of majority of the country's population depends on agriculture. The agriculture sector contributes only about 18 percent of the total Gross Domestic Product (GDP), with more than 60 percent population dependence, resulting in low per capita income in the farm sector. Consequently, there is a large disparity between the per capita income in the farm sector and the non-farm sector. Therefore, it is essential to deal with those issues which impact the income levels of farmers. The income levels are determined by the overall production, supported by reasonable levels of yield and prices realised by the farmers. Several constraints such as preponderance of small and marginal holdings accounting for about 82 percent of total holdings, imperfect market conditions and lack of backward and forward linkages affect the income levels of farmers adversely. Accordingly, an appropriate policy needs to be evolved to ensure that farming activity becomes more viable and the economic condition of farmers is improved on a sustainable basis.

Several steps have been taken in the past to strengthen agricultural research, education and extension, as well as to ensure timely and adequate availability of essential inputs such as seeds, fertilizer and electricity. Also, several major and minor irrigation projects have been implemented. An integrated programme of agricultural development was initiated in the early 1960s to improve productivity and to benefit farmers. Our scientists contributed by developing high yielding varieties/hybrids of crops. These were introduced to farmers who, in turn, adopted the new seeds and technology. As a result, a major breakthrough in productivity and production occurred in wheat in the late 1960s. Moreover, the new strategy laid emphasis on greater cropping intensity. Since then, the progress of agricultural production was steady and the growth rate in food production maintained a level above that of population growth. This came to be known as the "Green Revolution Era" that resulted in a situation of self sufficiency in foodgrains. This era was characterized by synergy among technology, services, public policy and farmers' entrepreneurship. However, during the past decade, the growth rate shows a declining trend; production and productivity have remained almost stagnant. A considerable effort, therefore, is required to reverse the deceleration and get agriculture moving forward once again.

The National Agriculture Policy approved by the Government of India during 2000, aimed to achieve annual growth of more than 4 percent in the agriculture sector on a sustainable basis, through the efficient use of natural resources and combination of other measures. However, the annual growth rate achieved during the Tenth Five Year Plan (2002-03 to 2006-07) averaged around 2.3 percent. On the other hand the non-farm sector has grown faster. Considering the high growth of GDP in the recent past, a major reorientation in the policy is necessary to make this growth more inclusive. The decline in agricultural growth coupled with declining profitability in the agriculture sector, in the face of rapid growth of non-farm sector, is one of the major concerns. The declining profitability is partially due to fluctuating world prices of agricultural commodities and the efforts to keep the domestic prices low to protect consumers' interest, which resulted in decline in the terms of trade for the farm sector. This, coupled with the stagnating and declining yield levels, resulted in low income to the farmers.

Another major problem confronting the rural areas in general and farm households in particular is the lack of employment opportunities. Efforts are needed to increase job opportunities in the farm sector through increased investment in irrigation, watershed development, wasteland development, land reclamation, etc. In addition, there has to be a greater focus on the accelerated development of the rural non-farm sector and development of clusters around towns/market centres. A growing farm sector, better rural infrastructure and connectivity, skill development, adequate power supply and easy availability of credit would help in the creation of more employment opportunities in the rural non-farm sector and, in turn, enhance the income of farm households.

In view of the above, there is a need to focus more on the economic well-being of the farmers, rather than just on production. Socio-economic well-being must be a prime consideration of agricultural policy, besides production and growth. The aim of the Policy is, therefore, to stimulate attitudes and actions which should result in assessing agricultural progress in terms of improvement in the income of farm families, not only to meet their consumption requirements but also to enhance their capacity to invest in farm related activities.

Recent Initiatives

Several significant initiatives have already been taken in recent years by the government to reverse the downward trend in agricultural production and to find sustainable solutions for strengthening the farmers' livelihood and income. Some of these important initiatives include: (i) Bharat Nirman; (ii) National Rural Employment Guarantee Programme; (iii) National Horticulture Mission; (iv) Expansion of Institutional Credit to Farmers; (v) Establishment of the National Bee Board; (vi) Establishment of the National Rain-fed Area Authority; (vii) Establishment of the National Fisheries Development Board (NFDB); (viii) Watershed Development and Micro Irrigation Programmes; (ix) Reforms in Agricultural Marketing and Development of Market Infrastructure; (x) Revitalisation of Cooperative Sector; (xi) Agri-business Development through Venture Capital Participation by the Small Farmer Agri-business Consortium; (xii) Reform and Support for Agriculture Extension Services; (xiii) National Rural Health Mission; (xiv) National Food Security Mission; (xv) *Rashtriya Krishi Vikas Yojana* to incentivise the states to invest more in agriculture; (xvi) Integrated Food Law; (xvii) Legislative Framework for Warehousing Development and Regulation; (xviii) Protection of Plant Varieties and Farmers' Rights (PPVFR) Act, 2001; (xix) National Bamboo Mission and (xx) Knowledge Connectivity through Common Service Centres (CSC) and IT initiatives.

Major Policy Goals

The major goals of the National Policy for Farmers are:

1. To improve economic viability of farming by substantially increasing the net income of farmers and to ensure that agricultural progress is measured by advances made in this income.
2. To protect and improve land, water, bio-diversity and genetic resources essential for sustained increase in the productivity, profitability and stability of major farming

systems by creating an economic stake in conservation.

3. To develop support services including provision for seeds, irrigation, power, machinery and implements, fertilizers and credit at affordable prices in adequate quantity for farmers.
4. To strengthen the bio-security of crops, farm animals, fish and forest trees for safeguarding the livelihood and income security of farmer families and the health and trade security of the nation.
5. To provide appropriate price and trade policy mechanisms to enhance farmers' income.
6. To provide for suitable risk management measures for adequate and timely compensation to farmers.
7. To complete the unfinished agenda in land reforms and to initiate comprehensive asset and aquarian reforms.
8. To mainstream the human and gender dimension in all farm policies and programmes.
9. To pay explicit attention to sustainable rural livelihoods.
10. To foster community-centred food, water and energy security systems in rural India and to ensure nutrition security at the level of every child, woman and man.
11. To introduce measures which can help attract and retain youths in farming and processing of farm products for higher value addition by making it intellectually stimulating and economically rewarding.
12. To make India a global outsourcing hub in the production and supply of the inputs needed for sustainable agriculture, products and processes developed through biotechnology and Information and Communication Technology (ICT).
13. To restructure the agricultural curriculum and pedagogic methodologies for enabling every farm and home science graduate to become an entrepreneur and to make agricultural education gender sensitive.
14. To develop and introduce a social security system for farmers.
15. To provide appropriate opportunities in adequate measure

for non-farm employment for the farm households.

Definition of Farmer: For the purpose of this Policy, the term *farmer* will refer to a person actively engaged in the economic and/or livelihood activity of growing crops and producing other primary agricultural commodities and will include all agricultural operational holders, cultivators, agricultural labourers, sharecroppers, tenants, poultry and livestock rearers, fishers, beekeepers, gardeners, pastoralists, non-corporate planters and planting labourers, as well as persons engaged in various farming-related occupations such as sericulture, vermiculture, and agro-forestry. The term will also include tribal families/persons engaged in shifting cultivation and in the collection, use and sale of minor and non-timber forest produce.

Asset Reforms to Empower Farmers

The purpose of asset reforms is to ensure that every farmer household in villages possesses and/or has access to productive assets like land, livestock, fishpond, homestead farm and/or income through an enterprise and or market-driven skills, so that the household income is increased substantially on a sustainable basis. This, in turn, would ensure nutrition and livelihood security and their education and health requirements.

Land: Considering the skewed ownership of land, it is necessary to strengthen implementation of laws relating to land reforms, with particular reference to tenancy laws, land leasing, distribution of ceiling surplus land and wasteland, providing adequate access to common property and wasteland resources and the consolidation of holdings. Following the conferment of land rights to women under the Hindu Succession (Amendment) Act, 2005, the provision of appropriate support services to women farmers has become urgent. Joint *pattas* for both homestead and agricultural land are essential for empowering women to access credit and other services.

The Land Acquisition Act would be reviewed with particular reference to the assessment of compensation. Prime farmland must be conserved for agriculture except under exceptional circumstances, provided that the agencies that are provided with agricultural land for non-agricultural projects should compensate for treatment and full development of equivalent degraded/wastelands elsewhere. Further, the commitment under the existing re-settlement policy of the Central/State governments would be fulfilled in letter and spirit. For non-agricultural purposes, as far as possible, land with low biological potential for farming would be earmarked and allocated. State governments would be advised to earmark lands with low biological potential such as uncultivable land, land affected by salinity, acidity, etc., for non-agricultural development activities, including industrial and construction activities.

Water: Non-availability of timely and adequate water for irrigation is now becoming a serious constraint in achieving higher productivity and stability of farming in many parts of the country. Therefore, assured irrigation is the need of the hour. Though the total rainfall in our country is satisfactory, its distribution is highly skewed. Therefore, rainwater harvesting and improving the efficiency of water use are important. It has been assessed that even a 10 percent increase in the present level of water-use efficiency in irrigation projects may help to provide life-saving irrigation to crops in large areas. Water-use efficiency can be further enhanced by generating synergy with seed varieties, nutrients (macro and micro) and farm implements. The concept of maximising yield and income per unit of water would be used in all crop production programmes. Water users' associations would be encouraged to gain expertise in maximising the benefits from the available water.

The majority of farmers depend on groundwater for irrigation. This resource, in which farmers may have invested their hard-earned savings, is being depleted and the water table is receding fast. Therefore, rainwater harvesting and aquifer recharge would be accorded priority for ensuring the stability and sustainability of water supply. Water quality also needs attention, since it often gets polluted at the source due to its over-exploitation and the indiscriminate use of fertilizers, pesticides and toxic chemicals.

Besides addressing the problems relating to adequacy and quality, equity in water distribution would be ensured. Water is a public resource, and not a private property. Therefore, priority would be given to evolve mechanisms for just and equitable access to water and to include local people in managing water resources. Women would be given a significant role as water users, both in access and management.

The following steps would be taken for augmentation of water availability and its efficient use:

- Rainwater harvesting and aquifer recharge would be given priority for ensuring the stability of supply. Necessary legislative measures to regulate and control the development and management of ground water would be taken up simultaneously.
- Existing wells and ponds would be renovated.
- Demand management through improved irrigation practises, including sprinkler and drip irrigation, and through *Pani Panchayats* or water users associations would be accorded a high priority.
- A water literacy movement would be launched and regulations would be put in place for the sustainable use of ground water.
- Integrated and coordinated development of surface and ground water resources and their conjunctive use would be envisaged right from the project planning stage and would form an integral part of project implementation.
- In water-scarce areas, the land-use system would emphasise the cultivation of crops of high value and requiring less water such as pulses and oilseeds.

  Symbiotic interaction and convergence of efforts would be

made through various initiatives of the central/state governments such as the National Rain-fed Area Authority, the National Horticulture Mission, the Technology Missions on Oilseeds and Pulses and the National Rural Employment Guarantee Programme to promote water-use efficiency and water conservation measures.

For drought-prone areas, a Drought Code would be introduced identifying the action needed to minimise the impact of adverse monsoons and to maximise the benefits of a good season. Similarly, in areas prone to heavy rainfall, a Flood Code would be introduced to mitigate distress, take care of the needs of the farmers immediately after floods and help convert the flood-free seasons into major agricultural production periods. For the arid areas, a Good Weather Code would be introduced for taking advantage of occasional heavy rainfall for strengthening the ecological infrastructure essential for sustainable livestock production, drinking water security and sand dune stabilization. The National Rain-fed Area Authority would provide technical and other support in this regard.

Livestock: Livestock, including poultry, is also one of the important sources of livelihood, contributing about one fourth of the agricultural GDP. These activities engage women in much larger proportion. The ownership of livestock is much more egalitarian since poor farmer families mostly own cattle, buffalo, sheep and goats. The major constraints experienced by farmers relate to breed, fodder, feed, healthcare and remunerative prices for their produce. There is a need to address these issues through an appropriate strategy. Further, to enhance the income of livestock owners, agri-clinics operated by veterinary and farm science graduates would be encouraged to improve productivity and overall efficiency of livestock. At the same time, crop-livestock mixed farming systems would be promoted, apart from encouraging production of organic manures and bio-fertilizers. Livestock insurance would also be revamped and made accessible to all farmers.

For poultry farming, the following steps would be taken:

- Quarantine and testing facilities for imported birds and vaccines at all ports of entry would be established and strengthened, since such safeguards are necessary for the health and survival of the poultry industry and for the protection of life and livelihood.
- Testing for safety and efficacy of imported poultry vaccines before they are allowed to be marketed, as is done in the case of human vaccines, would be made compulsory.
- Poultry rearing would be recognised as an agricultural activity and appropriate support would be provided to backyard poultry farmers to promote clusters or small holders' poultry estates.

Fisheries: Both coastal and inland fisheries provide employment and livelihood to millions of families. There is considerable scope for improving the income of fishermen families on an environmentally sustainable basis by encouraging scientific fish rearing, harvesting and processing. In the area of public policy, there is a need for well-planned aquarian reforms in order to provide landless labour families access to village ponds and other water bodies in the public domain for aquaculture.

The NFDB has been established by the Government of India to bring major activities relating to fisheries and aquaculture for focused attention and professional management. The guiding principles for the NFDB would be ecology, economics, gender equity, employment generation and advising the state governments to facilitate rational and equitable allocation of public water bodies and reservoirs for practicing modern aquaculture.

Following steps will be taken to promote modern aquaculture practices and value addition:

- "Fish for All" training and capacity building centres to impart training to fisher families and fisher-women in all aspects of the capture/culture/consumption chain, quality literacy for hygienic handling and other aspects.

- Provision of small dredgers for ensuring the efficiency of fish landing centres.
- Centralised services to support the decentralised capture and culture fisheries sectors.
- Inland aquaculture, including the culture of ornamental fish and air breathing fish, by providing necessary space in ponds and reservoirs.
- Artificial coral reefs to compensate for the loss of natural coral reefs to revive the fish catch.
- Raising bio-shields comprising Mangroves, Casuarina, Salicornia, Aatriplex and other halophytic plants, to safeguard the lives and livelihoods of coastal fisher and farm families in the event of cyclonic storms and seawater inundation during calamities like tsunamis.

A dynamic policy for the management and economic use of the Exclusive Economic Zone (EEZ) for a variety of economic activities, including fisheries, will be evolved and put in place with the assistance of NFDB.

Bio-resources: Bio-resources refer to the abundant wealth of flora and fauna, including soil micro flora and micro fauna, which, after land and water, form the third important natural resource available to farmers. Efforts will be made to conserve as well as enhance these resources and to ensure their sustainable use with equitable sharing of benefits. Two major legislations—the Protection of Plant Varieties and Farmers' Rights (PPVFR) Act, 2001 and the Biological Diversity Act, 2002—are in place to achieve some of the above aims. Implementation of these Acts would be strengthened and detailed guidelines would be developed in such a manner as to recognise the rights of farmers and the farming community. The National Gene and Biodiversity Fund would be used to recognise and reward contributions of farmers and to support revitalisation of *in situ* farm conservation traditions of such communities. A beginning has already been made by the government in establishing the Plant Genome Saviour Community Recognition Award for farmers.

For conservation and development of bio-resources, the following would be promoted:

- Documentation of traditional knowledge through community bio-diversity registers with the involvement of women, who hold much of this knowledge.
- Support to tribal and rural people for revitalising their *in situ* farm conservation traditions.
- Participatory breeding procedures involving scientists and local conservers for improving the productivity of land races.
- Genetic engineers working in public institutions to perform the role of pre-breeding, i.e. development of novel genetic combinations for important biological and economic traits, such as resistance to biotic and abiotic stresses, in participatory breeding programmes with farmers, to integrate genetic efficiency and genetic diversity in an effective manner.
- Genetic homogeneity enhances genetic vulnerability to pests and diseases. Therefore, pre-breeding and participatory breeding would be integrated to help insulate small farmers from the risks of pest and disease epidemics.
- Launch of literacy movement on genetic and legal aspects, in areas rich in agro-biodiversity such as the North East, Western and Eastern Ghats and the arid and semi-arid zones.
- Genome clubs will be encouraged in rural schools and colleges for imparting an understanding of the importance of genetic resource conservation.
- Literacy and awareness to help tribal and rural families understand the provisions of the PPVFR Act and the Biological Diversity Act with reference to their entitlements.
- Training of farm and tribal families in methods of preventing gene erosion.

- Conservation of coastal biodiversity, including coral reefs and sea grass beds and support to traditional methods of conservation.
- Organise and support herbal bio-valleys in the Western Ghats, Eastern Ghats, Vindhyas and Himalayan region for the conservation and sustainable use of medicinal plants. In such bio-valleys, farmers would be assisted through venture capital and other support to take to conservation, selection and multiplication of medicinal plants of value to health security.
- A nationwide programme will be taken up for the *ex situ* and *in situ* conservation of plant genetic resources at the field/farmer level. Farmer-level gene/seed banks would be set up in areas where traditional varieties are in danger of extinction. Seed exchange programmes undertaken by some state governments, should take due care to ensure that the traditional rice gene pool is not lost in this process.
- Participatory management of national parks, bio-sphere reserves and gene sanctuaries.

Animal Genetic Resources: A system of rewards and incentives will be developed to enable and motivate people to conserve their breeds under the Biological Diversity Act. The Biodiversity Fund would be used for such purposes. Livestock keepers' inherent rights to continue to use and develop their own breeding stock and breeding practises will be acknowledged and encouraged. The government will recognise these rights, acknowledge livestock keepers' contribution to the national economy, and adapt its policies and legal frameworks accordingly so as to protect the same and to pre-empt attempts to use the intellectual property system to obtain control over animal resources.

Apart from conserving genetic diversity and acknowledging the vital role of livestock keepers, there is a need to document the indigenous knowledge of pastoral communities about animal maintenance and breeding. Community-based conservation and development of indigenous livestock breeds and species will be encouraged, with a special focus on both hot and cold arid and semi-arid areas where genetic diversity and associated indigenous knowledge are particularly well developed. State farms may be used to promote in situ conservation of animal breeds. Grazing lands must be earmarked to enable the conservation of animal genetic resources. Documentation of special biological and economic traits would be done in the context of the new biological and nutritional needs or for other economic traits like hide/leather quality. Creation of disease-secure facilities for screening of germ plasm and selection of disease resistant varieties would be promoted.

Indian breeds of cattle and buffaloes are in demand in other countries. Animal science graduates, self-help groups (SHGs) and progressive livestock farmers would be encouraged and supported to maintain pedigree animals of these breeds for exploiting export opportunities. However, export of all biological material including animals would be in accordance with the provisions of the Biological Diversity Act.

Support Services

Science and Technology: Science and technology are the key drivers of change in farm operations and output. New technologies which can help enhance productivity per unit of land and water are needed for overcoming the prevailing technology fatigue. Frontier technologies like biotechnology, ICT, renewable energy technologies, space applications and nano-technology provide opportunities for launching an "Evergreen Revolution" capable of improving productivity on a sustainable basis. In order to ensure social inclusion in access to new technologies, public investment in socially relevant agricultural research should be stepped up under the umbrella of the National Agricultural Research System (NARS) comprising large numbers of Indian Council of Agricultural Research (ICAR) institutions, state agricultural universities, all-India coordinated research projects and national bureaus.

Non-governmental organizations (NGO) and private sector research and development (R&D) institutions would also be included under the NARS umbrella. NARS would be restructured so as to effectively address the problems faced by small and marginal farmers.

The research strategy should be pro-nature, pro-small farmer and gender sensitive. Community-managed seed villages and seed technology training centres are needed, with women playing the major role because of their traditional knowledge of seeds and seed management, especially in tribal communities. Scientific literacy and removal of doubts and fears about the risks and benefits associated with biotechnology and other new technologies can be achieved through farmers selected in each *panchayat* and provided with adequate training, so as to enable them to serve as farm science managers in their respective villages.

A National Bio-technology Regulatory Authority would be set up for ensuring the safe and responsible use of recombinant DNA technology or genetic engineering.

Need-based breeding of crop varieties would be stepped up such as processing quality fruits and vegetables.

High level multidisciplinary effort will be made for enhancing scientific inputs in organic farming that meet the needs of farmers. Integrated crop – livestock - fish production systems offer scope for the adoption of the principles and methods of organic farming.

In intensively cultivated mono-crop areas, crop diversification may be beneficial from the points of view of ecology, economics and employment generation. However, any advice on crop diversification must be accompanied by steps to ensure effective market support for the alternative crops. In planning for crop diversification, particularly from food to non-food crops, such as for the production of bio-fuels, the food security of the nation would be kept in view.

The intellectual property rights regime will make provisions for compulsory licensing of rights in the cases of research products and processes of value to farmers. In all cases of health and food security, social inclusion will be the guiding factor.

Conservation farming would be given priority in the heartland of the Green Revolution and turning the Indo-Gangetic Plains into a major food basket through an appropriate mix of technology services and public policies.

Agricultural Bio-security: Agricultural bio-security covering crops, trees and farm and aquatic animals is of great importance since it relates to both the work and income security of a majority of the population and the food and trade security of the nation. An integrated National Agricultural Bio-security System (NABS) covering crops, animal husbandry, fisheries, forestry and agriculturally relevant micro-organisms will be established with the following objectives:

- Safeguard the income and livelihood security of farmer families, as well as the food, health and trade security of the nation, through effective and integrated surveillance, vigilance, prevention, and control mechanisms designed to protect the productivity and safety of crops, farm animals, fishes and forest trees.
- Enhance national and local capacity in initiating proactive measures in monitoring, early warning, education, research, control and international cooperation.
- Introduce an integrated bio-security package comprising regulatory measures, education, improved sanitary and phyto-sanitary measures and social mobilisation.
- Organise an integrated national agricultural bio-security programme on a hub-and-spoke model, with effective home and regional quarantine facilities capable of insulating the major agro-ecological and farming system zones of the country from invasive alien species of pests, pathogens and weeds as well as from the introduction and release of Genetically Modified Organisms (GMOs).

Agro-meteorology: The nation has considerable capacity in short, medium and long range weather forecasting. Generic information about weather has to be translated into location-specific land-use advice, based on cropping patterns and water availability. The agro-meteorological advisories issued from time to time, would be used by Panchayat-level functionaries, trained to give appropriate land-use suggestions to farmers with the least possible time lag. For marine fisheries, data on wave heights and location of fish shoals available would be transmitted to the fishermen. Frontline technologies such as internet-FM/HAM radio/cellphone services would be very helpful to fishermen in this regard.

Climate Change: Climate change leading to adverse changes in temperature, precipitation and sea level is an emerging issue. Most experts agree that the impact of global warming has been in a worrisome measure, as evident from the melting of glaciers and Antarctic and Arctic ice caps. Coastal storms and cyclones are also increasing in frequency and intensity. Consequent to these changes, droughts and floods are likely to be more frequent. Although climate change is a product of unsustainable consumption of non-renewable forms of energy, the harmful impact of climate change will be felt more by those nations with limited resources and coping capacity. Proactive measures to reduce the vulnerability to climate change will be taken. Based on simulation models, contingency plans and alternative land use and water-use strategies will be developed for each major agro-climatic zone. In drought and flood-prone areas, experienced farmers would be trained as "Climate Managers" in the art of managing drought, flood and aberrant monsoons.

Inputs and Services

***Seeds*:** Good quality seeds and disease-free planting material, including in-vitro cultured propagules, are essential for crop productivity and security. Hybrids are now becoming available for many crops. In the case of new varieties, foundation seeds would be provided to grass root level seed growers and their groups such as cooperative societies and SHGs. Mutually beneficial farmer-seed company partnerships will be encouraged. Agricultural universities would be encouraged to organise courses on seed technology and business, and mainstream business principles in all applied courses. A national seed grid will be established to ensure supply of seeds across the country, as per the area specific requirement.

Soil Health: Soil health enhancement holds the key to raising farm productivity. Steps would be taken to ensure that each farmer is issued with a soil health passbook containing integrated information on the physics, chemistry and microbiology of farm soils with corresponding advisories. More laboratories to detect specific micronutrient deficiencies in soils will be established for this purpose. Soil organic matter would be increased by incorporating crop residues in the soil. Proper technical advice on the reclamation of wastelands and on improving their biological potential will be made available. Fertilizer pricing policies will be reviewed to promote balanced use of fertilizers. Agro-forestry will be encouraged for efficient nutrient cycling, nitrogen fixation, organic matter addition and for improving drainage. Appropriate production and marketing mechanisms for bio-fertilizers, organic manures etc. will be put in place to promote their use for improving/maintaining soil health.

Pesticides: The triple alliance of pests, pathogens and weeds is the cause of substantial crop losses every year. The development, introduction and diffusion of environmentally safe and effective pesticides will be given priority. There is a need for incorporating the use of chemical pesticides in an Integrated Pest Management (IPM) system. Suitable quality control, safety evaluation and other regulatory systems would be strengthened. The sale of spurious and substandard pesticides would be prevented and bio-pesticides would be promoted.

Implements: Farmers need region and crop specific machines and implements for timely sowing, management of weeds and improving their post-harvest operations. Women especially need woman-friendly implements/tools which can reduce drudgery, save time, enhance output and can be handled comfortably. Agri-entrepreneurs including farm graduates and progressive farmers would be encouraged to provide implements and tools, machinery, tractors and other farm implements on a custom-hire basis.

***Vaccines and Sero-diagnostics*:** Major gaps in the presently available facilities would be filled in the case of important animal diseases. Bio-technology research in the area of vaccine development would be stepped up, encouraging public private partnerships.

***Fish Seed and Feed*:** Good quality and disease-free fish seed holds the key to successful inland aquaculture. Progressive fishermen and their groups such as SHGs would be trained in induced fish seed breeding, production and availability of seed and fish seed feed at affordable prices, with appropriate technical assistance from the NFDB and other agencies.

***Animal Feed*:** Inadequate nutrition is the primary cause of low milk yield in dairy animals. Both conventional and non-conventional approaches for conversion of cellulosic wastes into good animal feed through appropriate treatment and enrichment, planting of nutrition rich fodder plants and dissemination of technologies would be encouraged.

***Other Essential Support Services*:** These include establishing genetic evaluation systems for indigenous breeds as well as crosses, so that selection can lead to genetic improvement of production characteristics; upgrading of breed through artificial insemination; cross- breeding suited to the farmers' resources; and improved processing and marketing. The livestock sector has to become sanitary and phyto-sanitary compliant. A cadre of paraveterinarians would be trained to set up support service centres for the farmers, to promote early detection and treatment of diseases and other veterinary services.

Support Services for Women Empowerment: For capacity

building and livelihood, women working in the farms need appropriate support services like crèches, child care centres, nutrition, health and training, etc. For funding such activities, existing schemes of the Ministry of Panchayati Raj, the Ministry of Rural Development and the Ministry of Agriculture would be augmented and utilised and new schemes would be introduced, if required.

Credit and Insurance: Improvement in the outreach and efficiency of the rural banking system is the need of the hour. Towards this end, the financial services would be galvanised for timely, adequate and easy reach to the farmers at reasonable interest rates. The banking system would endeavour to meet the large credit potential needed to raise agriculture to higher thresholds and for the growth of rural and agribusiness enterprises and employment, and would take steps to achieve financial inclusion.

The Government of India has already put in place an agriculture credit policy to improve access of farmers to institutional credit. Steps would be taken for extensive coverage of farmers under the *Kisan* Credit Card Scheme. Micro credit and micro insurance will be promoted as an effective tool for encouraging production and reducing risk. Credit cooperatives have an important position and role in the rural financial system and priority would be given to reforms and revamping of cooperative credit institutions as per the recommendations of the Vaidyanathan Committee. Credit counselling centres would be established where severely indebted farmers can be provided a debt rescue package/ rescheduling to save them from a debt trap. National Bank for Agriculture and Rural Development (NABARD), as the leader of agriculture and rural credit, should facilitate convergence between credit availability and credit absorptive capacity of the farmers and other rural borrowers and an efficient credit delivery system. NABARD should actively involve itself in institution building and provide backup support through research and development initiatives. NABARD should function like a national bank for farmers.

Since agriculture is a high-risk economic activity, farmers need user-friendly insurance instruments covering production, right from sowing to post-harvest operations. The insurance should also cover the market risks for all crops, in order to insulate the farmers from financial distress and in the process make agriculture financially viable. Steps would be taken to revamp the National Agricultural Insurance Scheme to make it more farmer-friendly.

There is also a need for credit and insurance literacy in villages. *Gyan Chaupals* (village knowledge centres) can help in this task. Awareness on credit and insurance issues among farmers would be promoted.

Women need special attention in credit access because of their lack of land title/collateral. *Kisan* credit cards would be issued to women speedily with joint *pattas* for homestead/agricultural land. For the cases without joint *pattas*, indemnity bonds/guarantees from husband, and relatives would be considered by the banks for extending credit and *kisan* credit cards to the women farmers.

Cooperatives: Cooperatives have an important role to play in banking, input supply, marketing, agro-processing and other agribusinesses to protect farmers from the vagaries of existing imperfections in the supply of inputs, production, value addition and marketing. Cooperatives should function as economic enterprises and not as an extended arm of the state. They require an entrepreneurial approach, competitive edge through suitable enterprise focus and strategic alliances with private and public sector units. Appropriate mechanisms would be put in place so that farmers have greater control of the market channels and improve profit opportunities through cooperatives and SHGs.

With economic liberalisation and market competitiveness, cooperatives would require much larger capital and other financial resources. Changes in the legal framework and regulatory system would help gain greater access to capital/financial resources. The policy and legal framework under which cooperatives are functioning would be reviewed so as to create an enabling environment for them to attain autonomy and run their operations in a business-like manner, subject to provisions of law. The management of the cooperatives needs to be made professionally competent, with clear demarcation of functions of the elected members and the managers. The audit and accounting systems would be improved and made transparent so as to give greater confidence to all the members of cooperatives.

Extension, Training and Knowledge Connectivity: The gap between scientific know-how and field level do-how has been widening in the recent years. This knowledge deficit would be overcome speedily to enhance farm productivity and profitability. *Krishi Vigyan Kendras* (KVKs) would take up training and lab-to-land demonstrations in the area of post-harvest technology, agro-processing and value addition to primary products to provide skilled jobs in villages. State governments would be supported for strengthening the extension machinery through retraining and retooling of existing extension personnel and for promoting farmer to farmer learning by setting up farm schools in the fields of outstanding/progressive farmers. The farm schools with linkages to KVKs can speed up the process of technological upgradation of crop and animal husbandry, fisheries and agro-forestry. Efforts will be made to bring farmers, processors, retailers and other stakeholders together to support modern agricultural practices. Convergence of extension efforts especially at the district level and below would be ensured.

The potential of ICT would be harnessed by establishing *Gyan Chaupals* in villages. Further, the common service centres of the Department of Information Technology, Government of India and those set up by the state governments and private initiative programmes will be evolved for inclusive and broad-based development. Thus, the structure of the ICT-based knowledge system would, inter alia, include setting up of such village centres. Last-mile and last-person connectivity would be facilitated with the help of technologies such as broadband internet, community radio, or internet-mobile phone synergies.

Empowering farmers with the right information at the right time and place is essential for improving the efficiency and viability of small and marginal holdings. Mass media, particularly the radio, television and local language newspapers, will be used to play an important role in this regard.

Social Security: Coverage of farmers, particularly small and marginal farmers and landless agricultural workers, under a comprehensive national social security scheme is essential for ensuring livelihood security. The government would, therefore, take necessary steps to put in place an appropriate social security scheme.

Agricultural Prices, Marketing and Trade: Assured and remunerative marketing opportunities hold the key to continued progress in enhancing farm productivity and profitability. Several significant market reforms have already been initiated by the Central and the State governments. These reforms provide more options to farmers for selling their produce, allowing the private sector, including cooperatives, to develop markets, promote direct sales to consumers, processors and retail chain suppliers/exporters and remove scope for corruption and harassment. The following steps would be taken:

- The Minimum Support Price (MSP) mechanism would be implemented effectively across the country.
- The Market Intervention Scheme (MIS) would be strengthened to respond speedily to exigencies especially in the case of sensitive crops in the rain-fed areas.
- The establishment of community foodgrains banks would be promoted to help in the marketing of underutilised crops and thereby generate an economic stake in the conservation of agro-biodiversity.

- The food security basket will be enlarged by storing and selling nutritious millets such as bajra, jowar and ragi and other crops through the network of the Public Distribution Systems (PDS).
- Efforts will be made to develop a single national market by relaxing internal restrictions. All controls and regulations hindering increase in farmers' income will be reviewed and abolished.
- Terminal markets for agriculture would be developed in public-private partnership mode to provide better market access to farmers with better price realisation in a transparent trading environment with suitable backward linkages to give technical backstopping services needed for quality and demand driven production.
- The role of the Agriculture Produce Market Committees and State Agriculture Marketing Boards would be transformed from mere regulatory focus to promotion of grading, branding, packaging and development of markets for local produce.

Farmers require authentic advice based on meteorological, marketing and management information for land-use decisions and investments. Infrastructure support would be put in place to minimise post-harvest losses and enable agro-processing and value-addition at the village level itself to increase employment and income. Farmers' organisations and other entities like cooperatives and small farmers' estates would be encouraged so that farmers can get a fair deal and enjoy the economies of scale. Producer groups and cooperatives will be encouraged to promote agro-processing industry. Constraints would be removed for improving the negotiability of warehouse receipts.

The trade policies in agriculture would aim at protecting the livelihood of farmer families and fostering their economic well-being. Effective livelihood security for farmers would be put in place. Quality and trade literacy programmes would be launched across the country. Appropriate measures would be introduced to mitigate price risks and enable the stakeholders, particularly the farmers, to hedge their risk.

Farmers' associations and SHGs would be supported to export on competitive terms by spreading awareness of the opportunities available for external agricultural trade and for value addition. The agri-export zones would be further strengthened to become places where farmers will get the best possible price for their produce.

The twin goals of ensuring justice to farmers in terms of a remunerative price for their produce and to consumers in terms of a fair and affordable price for staples (as farmers are also consumers) would be achieved through the following integrated strategy:

- The government, while taking decisions on MSP, would ensure that the farmers' interests in receiving remunerative prices for their produce are adequately safeguarded.
- Stable and efficient market environment, including effective implementation of MSP in rain-fed areas, would add to the improvement in productivity and income in dry land farming.

The terms of reference and status of the Commission for Agricultural Costs and Prices (CACP) would be reviewed to make the MSP regime more effective.

Integration with Processing and Value Chain: It is well-recognized that value addition to agricultural produce increases income levels of the farmers. Therefore, it is necessary to integrate their production with processing and other value chain activities. However, the post-harvest infrastructure in the country is grossly inadequate, resulting in huge inefficiencies and wastages. The percentage of post harvest losses at various stages of storage, grading, packing and marketing due to improper handling and transportation, vary greatly, depending on the nature of crop and climatic conditions. The food processing industry has the potential to provide long-term economic sustainability to the farmers by adding value to their produce, improving post-harvest management linking the farmers to the market and promoting demand-driven farming. Further, the food processing activities also need modernisation. Hence, effective steps would be taken to facilitate and stimulate this process, modernise the food processing sector and enhance its competitiveness.

Curriculum Reform: Agricultural/Animal Sciences Universities: The motto of these universities would be to groom "every scholar/student as an entrepreneur". This will call for integrating business management principles with major applied courses. The agriculture/farm universities would also restructure their curricula in a manner that the gender roles in farming are recognised and they are equally empowered technologically. There is a need to restructure the curricula to focus on nutrition, post-harvest technology, quality and safety standards. The farm universities would be reoriented to give emphasis on entrepreneurship and capacity building on quality including sanitary and phytosanitary measures and Codex Alimentarius standards of food safety.

The mandate of the ICAR would be expanded to provide for registration and accreditation to the farm graduates as registered farm practitioners to provide quality service to the farmers. Centres of excellence in agriculture (crop, animal husbandry, fishery and forestry) modelled after the Indian Institutes of Technology (IITs) and Indian Institutes of Management (IIMs) would be developed to enhance the competitiveness of the graduates. Taking into account the new opportunities opened up by ICT, pedagogic methodology for promoting a learning revolution among students would be adopted. University centres would also establish employment and business advisory services and promotion centres and special one-stop windows for generating awareness of self-employment opportunities.

Special Categories of Farmers

Tribal Farmers: Tribal farmers are among the most disadvantaged category of farmers. A majority of tribal communities across the country are dependent on forests and animal husbandry for their livelihoods. These include cultivation (shifting cultivation in many cases), collection of fuel, fodder and a range of non-timber forest produce. Uplifting of their economic conditions would be a priority for the government. For this purpose, the following actions would be initiated:

- Updating land records in areas inhabited by tribal farmers.
- Strengthening the institutional structure to enable a more participatory process in decision making.
- Easy access to institutional credit to all tribal farmers and adequate provision of *kisan* credit cards for them.
- Documentation of the traditional crops and knowledge of the tribal farmers and creating an economic stake in conservation of such crops.
- Provision of appropriate technology and extension services and relaxed criteria for providing inputs such as water, fertilizers, seeds, etc. for tribal areas.

Pastoralists: The following steps would be initiated to ensure better livelihood opportunities for pastoralists:

- Restoration of traditional grazing rights and camping rights in respect of forest areas and in those areas earmarked for grazing purpose in village common lands.
- Formalising entitlements (including issue of permanent grazing cards) for traditional pastoralists/herders maintaining native animal breeds to enable free access to notified or demarcated grazing sites and migration routes.
- Grazing land and drinking water sources for livestock will be conserved and expanded to the extent feasible.
- In-depth documentation and characterisation of indigenous livestock breeds and preservation thereof would be carried out to recognise and protect the intellectual property rights of the local communities/individuals conserving these livestock breeds.
- Pastoralists would be involved in all local natural resource management programmes, including village forest

committees and joint forest management.

Other Categories: Besides the above two groups, there are several small groups with distinct and special needs such as small plantation farmers, island farmers, urban farmers and organic farmers.

Plantation Farmers: A large number of small farmers are engaged in the cultivation of plantation crops like tea, coffee, rubber, cardamom, pepper and vanilla. Price fluctuation and competition from products imported from abroad are among the major problems facing them. Therefore, the Price Stabilisation Fund for plantation crops will help them to be insulated from the vagaries of the market.

Island Farmers: The farming and fisher families in the Andaman and Nicobar Islands, Lakshadweep group of islands and other islands need special attention including technology, training, techno infrastructure and trade. Island agriculture also has the problem of transport costs, particularly for perishable commodities like fish which may have to be sold in the mainland. The ancient tribes of the Andaman and Nicobar Islands have rich traditional knowledge and wisdom. Steps would be taken to recognise and reward their indigenous knowledge in the areas of biodiversity conservation and traditional healthcare. Horticulture development programmes would be taken up in islands wherever feasible. The creation of mangrove and non-mangrove based bio-shields would be initiated to safeguard the lives and livelihoods of island populations in the event of the rise of sea levels due to global warming and calamities like tsunamis.

Urban Farmers: In urban areas, home gardens and nurseries would be encouraged. Nutrition gardens would be supported to supplement the nutritional requirement.

Special Categories of Farming

Organic Farming: The organic farming movement in India suffers from a lack of adequate institutional support in the areas of research, extension, certification and marketing and it requires more scientific support than chemical farming. KVKs would be equipped to provide training in organic agriculture. Internationally accepted certification procedures would be strengthened to make them farmer-friendly and affordable. Organic farming zones would be identified, like the hill areas, islands where chemical fertilizer use is low, and for medicinal plants where the use of chemical pesticides and fertilizers is not advisable. Organic farming would also be encouraged in selected rain-fed areas backed up by required supports, including marketing. It would also be supported through contract farming. Food safety and quality specifications should conform to the Codex Alimentarius standards since there are occasional reports of heavy metals being present in organic foods. Farmers engaged in organic farming should be linked to niche markets where they will obtain a premium price. Progressive farmers and farm graduates would be supported for establishing agri-clinics and agri-business centres for organic farming. Bio-fertilizers, organic manures and bio-pesticides would be treated at par with the chemical fertilizers for support and promotion.

Green Agriculture: Green agriculture involving integrated pest management, integrated nutrient supply and integrated natural resources management is recognized as the pathway to an "Evergreen Revolution". Unlike organic farming, green agriculture permits the safe and minimal use of mineral fertilizers and chemical pesticides, as well as crop varieties developed by genetic modification. Green agriculture products would be encouraged with distinct labelling and certification as in the case of organic farming.

Genetically Modified (GM) Crops: There is a need to assess the risks and benefits associated with GM crops in a credible and transparent manner. Priority would be given for genetic modification to incorporate genes which can help impart resistance to drought, salinity and other stresses. Water-use efficiency as well as improvement of both nutritive and processing quality would also be accorded priority in the research agenda. Training and awareness in agronomic management procedures in respect of GM crop varieties would be introduced.

Protected (Greenhouse) Agriculture: With the rapid growth of horticulture, there is an opportunity for greenhouse cultivation of vegetables, fruits and flowers supported by economic methods of water and fertilizer use such as fertigation (the application of nutrients through irrigation systems). Support would be given to farm and home science graduates and other entrepreneurs to undertake green house horticulture under the agribusiness programme and horticulture development programmes. Low-cost green houses, along with micro-irrigation and fertigation techniques, would be popularised in areas where evaporation exceeds precipitation during several months in a year. Suitable support would be given for such technologies that would increase income in water deficit areas.

Special Regions

Distress Hotspots: Several steps have been taken by the government to mitigate agrarian distress experienced in some parts of the country. It is important that special attention is paid to such areas to reduce risk and promote inputs for sustainable agricultural practices. In such areas, knowledge connectivity, social support systems and marketing infrastructure will be strengthened. Convergence of the benefits under different schemes would also be encouraged.

Mega Bio-diversity Areas: Local communities would be involved in the conservation of mega bio-diversity areas such as Western and Eastern Ghats, eastern Himalayan region, other North East and tribal areas. These communities would be given an incentive and their contributions would be appropriately recognised. Steps would also be taken to enable local communities in mega bio-diversity areas to convert bio-resources into economic wealth in a sustainable manner.

Farmers of the Future

The slow growth of opportunities in the non-farm employment sector has led to the proliferation of tiny and economically non-viable holdings. Increase in small farm productivity and creating multiple livelihood opportunities through crop-livestock integrated farming systems as well as agro-processing would be supported for increasing farmers' incomes. Methods of achieving economies of scale by small and marginal farmers through the farming groups would be popularised to enhance yield and income. Women would be encouraged to participate in such group activities. The following are some of the methods which may benefit the small and marginal farmers for gaining efficiency and economies of scale in their farming operations.

- **Cooperative Farming and Service Cooperatives:** These have been successful in the case of the dairy industry. Marketing cooperatives are successful since members cooperate on the basis of enlightened self-interest. Other forms of service cooperatives are slowly emerging, but the pace needs to be accelerated. Small farmers' cooperatives would, therefore, be encouraged and supported to take up activities such as processing, value addition and marketing of agro products.
- **Group Farming by SHGs:** SHGs for supporting micro-enterprises operated by women with the help of micro credit have been highly successful. With the growing diminution in the size of operational holdings, it will be useful to promote SHGs and commodity-based farmers' organisations at the production end of the farming enterprise by encouraging groups for consolidating farming activities. This will be particularly helpful in the case of "Green Agriculture" involving integrated pest management, integrated nutrient supply, scientific water management and improved post-harvest technology. SHGs would also be promoted for improving access to credit.
- **Small Holders' Estates:** The formation of small holders'

estates, for example, in cotton, horticulture, medicinal plants, poultry and aquaculture will help promote group cooperation among farmers living in a village or watershed, or the command areas of irrigation projects. Improving productivity, reducing the cost of production and entering into marketing contracts with textile mills, food processing industries, pharmaceutical companies, and fish marketing agencies will be some of the advantages. Such small holders' estates can also manufacture products under brand names and enhance income security so that group insurance becomes feasible. Agri-clinics and agri-business centres would be linked to such estates.

- **Contract Farming:** Symbiotic contracts which confer benefits to both producers and purchasers would be encouraged for ensuring assured and remunerative marketing opportunities to the farmers. Contract cultivation based on a well-defined code of conduct will be helpful to small producers in getting good quality inputs, fair prices and prompt payment for their produce. A code of conduct for contract farming or a model contract will be developed to meet the commodity-specific requirements. Farmers will not be alienated from their land under any circumstances under contract farming. State governments may set up a monitoring committee involving farmers to encourage a farmer-friendly contract farming system.
- **Farmers' Companies:** Private limited companies, registered under the Companies (Amendment) Act, 2002, are now coming into existence in the area of seed production, the production of bio-fertilizers, bio-pesticides and other forms of biological software essential for sustainable agriculture. Small farmers and the SHGs would be associated in such companies as stakeholders and not just as shareholders.
- **State Farms:** State farms inter alia, would also be used for developing Living Heritage Gene Banks of the germ plasm of local crops and breeds of cattle, sheep and poultry in

order to preserve the plant and animal genetic wealth.

Attracting Youth

Educated youths would be helped and supported for setting up agri-clinics and production-cum-processing centres to undertake outsourcing jobs both from within and outside the country. In order to attract youths to the agriculture sector, a number of vocational training courses in different aspects of agricultural and allied activities including value addition and processing of agro-products shall be introduced and recognised. The KVKs, institutions of state governments/ICAR and competent private institutions may also provide such recognised vocational training to the youth. Trained persons would be supported through various schemes to launch self-employment ventures for value addition to improve productivity and income of the farmers.

Other Policy Measures

The following policy measures will also ensure well-being and livelihood security of the farmers:

- Existing state land use boards would be revamped and linked to district-level land-use committees to be constituted by the state government so that they can provide quality and proactive advice to farmers on land use. These are expected to emerge as virtual organisations with the capacity to link land-use decisions with ecological, meteorological and marketing factors on a location and season-specific basis.
- Major thrust would be given to the development of floriculture, roots and tuber crops, aromatic and medicinal plants, bee-keeping and sericulture. Appropriate market linkages would also be provided so that such efforts would improve employment and income opportunities in rural areas.
- Market Intervention Scheme would continue and would be strengthened. Suitable measures would be taken for

expeditious processing and sanctions.

- A well-defined food security policy with home-grown foodgrains is important for eradicating rural poverty and malnutrition. In order to strengthen and regularly monitor food security issues, the government would constitute a Cabinet Committee on Food Security.
- Farm families would also be protected from becoming victims of HIV/AIDS and tuberculosis (TB). Anti-retroviral drugs would be made available in villages free of cost. The approach to the treatment of farmers who are victims of serious diseases like HIV/AIDS and TB in villages would take into account access of the patient to food and nutrition to assist recovery and restoration of productive life.
- There is a need for a rural non-farm employment initiative for farm households. The initiative would bring all rural non-farm employment programmes together in order to generate convergence and synergy among them. A consortium approach involving the different agencies concerned would be adopted.
- Along with production growth rates, income growth rates would also be measured and published. Improving income and livelihood of farmers would be incorporated as a primary objective of all schemes implemented in the agriculture sector.
- Article 243 G of the Constitution (73rd Amendment) Act, 1982, entrusts *panchayats* with the responsibility for agriculture including agricultural extension. If the problems faced by farmers are disaggregated and dealt with by the *gram sabhas* and *panchayats*, location-specific problems can be attended to speedily and effectively. In order to implement various schemes and programmes at the grass roots level, the Government of India would support state governments for devolution of functions and functionaries for empowering the *panchayats*. Steps would be taken to strengthen and accord centrality to *panchayats*

in addressing the problems of farmers.

- Mass media (conventional, electronic and the internet) constitute an important pillar of our democratic system of governance. In order to assist the mass media with timely and scientifically accurate information on issues relating to agriculture, food security and farmers welfare, regional media resource centres would be set up, associating farmers, media representatives and scientists, including extension personnel.
- There is a need for integrated approach for rural energy where renewable energy systems/devices, bio-fuels would be promoted apart from improving the rural infrastructure, including rural electrification, which is essential for agricultural development.

Operationalisation of the Policy

The National Policy for Farmers will be adapted and operationalised to suit the local needs in different states and union territories. State governments would be supported to convert national goals into local action points by preparing operational plans for implementing this policy, taking into account the agro-climatic and other local conditions. Such operational plans will be prepared at the district level by a multi-disciplinary professional group and integrated at the state level. States would be encouraged to have an effective mechanism for continuous feedback from the farmers to ensure that measures taken under the policy address the problems of the farmers.

The Department of Agriculture and Cooperation would constitute an Inter-Ministerial Committee to suggest a plan of action for operationalisation of this policy. Appropriate mechanisms and guidelines would be evolved by concerned Ministries/Departments to implement this policy.

The Agriculture Coordination Committee under the chairmanship of the Prime Minister would oversee and coordinate the integrated implementation of this policy.

Index

Index